ALIENS FROM ABOVE

ALIENS FROM ABOVE

the last in line

First edition

Michael J. Orrell

ALIENS FROM ABOVE: THE LAST IN LINE

Self-published via CreateSpace
www.createspace.com
(First edition, May 2017)

This is a work of part fiction and part non-fiction. Any similarity between characters and situations within its pages and places or persons, living or dead, is unintentional and co-incidental. Some names and identifying details have been changed to protect the privacy of individuals.

ISBN: 1546458395
ISBN 13: 9781546458395

About the Cover Art

'Spirit Valley'

This book's cover page, titled 'Spirit Valley', is a hand drawn illustration by the author himself using acrylic paint. Mike was inspired to create this image, upon his discovery of ten UFOs hovering across the backdrop of San Diego's popular El Cajon Mountain. The American Indian portrayed within 'Spirit Valley', referred to within Mike's writings as 'Whakan', is of his avatar self; contemplating the 'Sky Brothers' and their spiritual mission to help mankind.

About the Author

Michael Orrell is a native San Diegan whose work is inspired through all the scenic splendors that San Diego has on offer. He is an avid UFOlogist, photographer, artist and retired graphic designer, who photographed ten daylight UFOs in July of 1990, while hiking through San Diego's Inaja Memorial Park woodlands with two friends.

Upon having his photograph professionally enlarged, Mike discovered one object exactly matched the acorn-shape of the famous Kecksberg UFO that had crashed in Pennsylvania on December 9, 1965. Mike's photograph displayed a strange, spike-like projection, which he later found in numerous other UFO photos, including that of Astronaut James McDivitts. This anomaly would eventually turn out to be known as the *Rosetta Stone*, which successfully links UFOs from one to another, along with countless ancient artifacts across numerous civilizations.

Mike's discovery has been featured in every newspaper in San Diego, as well as a CBS special on its affiliate, *KFMB-TV Channel 8*. The *Los Angeles Times* broke the story and labeled the evidence on Mike's not-for-profit website and *YouTube* video as 'unsettling'.

Mike does not claim to be an expert on UFOs, but he is a UFO enthusiast who is interested in the search for truth relating to aliens and UFOs, which may help to usher in the next phase of human evolution.

Mike welcomes interview opportunities and discussion, and can be contacted by email through orre@cox.net.

Acknowledgement

I am my father's son and am most grateful to him for providing outstanding leadership for the first 13 years of my life before he crossed over. He led by example, giving his three sons insight into living productive and meaningful lives. Jim was a great humanist and taught his sons, by example, to have compassion for all living things. My father accomplished a great deal before his death that occurred due to injuries sustained in a 1971 dune buggy wreck in Cantamar, Mexico, at the young age of 38 years. It was because of his death, I was provoked to search for answers concerning the mysteries of our existence, which I later found through the study of metaphysics.

I also want to thank Dennis, my little league baseball coach, who stepped in after my Dad's death and introduced me to several fine metaphysics authors, such as T. Lobsang Rampa, Carlos Castanada and Jane Roberts, all of whom I offer my gratitude in memoriam. Without their combined insights into the mechanics of our reality, this book could not have been written and the UFO photo I took in 1990 would probably have been discarded as a negative glitch, or a flock of birds as friends suggested.

I am indebted to all who covered my breakthrough discovery on television and in print. Special thanks go to the *Los Angeles Times*,

Beach & Bay Press, The Daily Californian, Good Times Magazine, The Examiner, SLAMM: San Diego's Lifestyle and Music Magazine, and the region's largest newspaper, *The San Diego Union-Tribune,* who featured my story on their front page in early July of 2011. I remain grateful to the many courageous writers, editors, publishers and photo-journalists associated with these publications, as they helped to distribute my work.

Numerous publications and television shows aided my research by providing substantial corroborating evidence that confirmed I had indeed stumbled upon a lost and sacred pattern proving the existence of extraterrestrials within our human history as presented within the pages of this book through the successful linking UFOs and countless ancient artifacts to each other.

I humbly acknowledge *National Geographic*; *Smithsonian* magazine; *TIME-LIFE Books*; *Unsolved Mysteries*; Leonard Nimoy and his hosting of *In Search of...*; Arthur C. Clarke's *World of Strange Power*, Marilyn Bridges and her book *Markings: Sacred Landscapes from the Air*; along with the television show *Secrets and Mysteries* that aired a pivotal piece of evidence, being a lost UFO photo taken by astronaut James McDivitt aboard his *Gemini IV* flight. I also convey my enduring thanks to countless books and TV shows who contributed to expanding my knowledge toward this project.

Appreciation goes to *CBS News 8* and their affiliate *KFMB-TV Channel 8* for their superior film work in San Diego. Not only did they come to my home and interview me for a 2006 segment, which was heavily advertised through their network, they also accompanied me to Inaja Memorial Park and Scenic Spot 7, where I originally took my *Famous Inaja UFO Photo*. They were responsible for having my 35mm negative examined by experts at Chrome Photo Lab, who authenticated the negative as untouched and the objects exhibited within as unidentifiable. Eight years on, my *KFMB-TV*

Channel 8 interview was awarded by winning first place at the 4th Annual *McMinnville UFO Film Festival.*

Particular thanks to all my friends and family for their ongoing support, as well as to the Universe for sparing my life on numerous occasions, so I could find the key that opens doors to a very bright future for mankind. Such a prospect offers interplanetary contact and trade, all of which would mean an end to disease, starvation and homelessness, as Earth becomes the destination of choice for visiting beings from every corner of this Universe, and beyond.

Last, but not least, I would like to thank my editor, Janelle, for being a huge 'pain in my butt', yet the driving force behind the finishing of my manuscript, having since become this book.

Dedication

I dedicate this book with love and affection, to my beloved father James, whose life was given to God too early for those he left behind. Loved in life and beyond, a man of great integrity who adored his family and whom I continue to admire, and respect. I shall forever hold a unique and spiritual connection with my Dad, as I strive to devote my own life's work toward all that he cherished and treasured.

Contents

1

Dreams have Meanings

The cool ocean breeze slowly worked its way up the winding tributary, eventually reaching the rugged and sun-tanned face of Whakan; a strapping young American Indian brave of the ancient Kumeyaay tribe. He had planned his hunt to include a rest break by his favorite getaway that held views across the picturesque natural springs, west from his village encamp-ment in the valley.

This was a favorite spot for Whakan and one where he had enjoyed numerous afternoon get-togethers with his raven haired beauty, Johona, who came from the neighboring Cocopa tribe. Laying his hunting gear in the usual spot, located in-between two nearby small boulders at the edge of the bubbling brook, Whakan at last stretched out his muscular legs until they were knee deep in the pool's cooling waters.

A nearby mockingbird greeted Whakan with a colorful recital of the best songs in his neighborhood and then responded joyfully to Whakan's comical imitation of his performance.

"Such a magical spot this is," he thought, as the sun glinted across the trickling stream, whose mesmerizing clatter always calmed whatever troubles Whakan seemed to be dealing with. Aside from the two disgruntled local girls, who he now refused to date upon meeting Johona, his tribe also counted heavily on Whakan's unmatched skill at flint knapping. The numerous arrow points and axeheads he rapidly produced with superior craftsmanship and precision, were highly sought-after trade items at local monthly native Indian trade and tribal gatherings. He had discovered a secret location at the foot of a nearby majestic mountain where he often found good nodules of raw flint used to make dozens of knives, scrapers, and perforators; all readily sought at such events.

As Whakan relaxed by the gurgling brook, he pondered on his mystical experience, which had occurred two weeks prior, where he met and communicated with his spirit guide when participating in his tribe's sacred Peyote ceremony. It was his third exploration into this separate reality and on each occasion, it strengthened his resolve to find 'truth and loyalty', being was his tribes' religion. It was indeed wise advice given to him by his elder mentor, to purify his mind before the ceremony, as one must practice clarity of purpose before venturing into this spiritual realm that has iniquitous places and entities to be avoided.

A smile emerged on Whakan's lips, as he lay his head down on the lush grass. He was grateful to the Universe for all his many blessings. He valued his relationship with 'all that is' and eagerly looked forward to more life lessons from his elder mentor. His last thought before drifting off to sleep was of Johona's perfect lips, gently kissing him from above, while her long, flowing hair teasingly rolled across his brawny chest.

A loud crashing in the brush came from behind, startling Whakan awake and breaking the delicious spell that Johona had cast over him in his dream. Secretly, he suspected she was a sorcerer's apprentice, just as he also thought of himself. Quickly gathering his wits, he noiselessly went into his stalking mode, notching an arrow point in his bow, ready to attack. Moving slowly around the brush and boulder laying in his way, he searched for the prey that would become the evening meal for his hungry tribe, who eagerly awaited his return down in the valley. What he saw; however, shocked him so much that the grip on his bow loosened and he clumsily dropped it, along with an arrow, which noisily clattered upon the ground.

Standing in the clearing before him stood an all-white deer, the likes of which Whakan had not previously seen. As the ghostly white apparition prepared to bolt, he gently cooed to this deer, sending a hard and slow blink her way. Indeed, he had learned well the ways of animals and considered himself their friend.

Tribal elders had often discussed the meaning of the coming of this sacred albino creature, having decided it was a manifestation of the Spirit Father himself, needing to be revered and never harmed, in any way. Whakan believed this to be the very same albino deer often spotted by the tribes' women on numerous occasions as they pounded acorns into a mash by the stream on their beloved grinding stones. It was these same blue-granite colorful rock slabs, running all the way into the wide stream, which had become a favored gathering place for the many native Indian tribes who regularly met at this watercourse.

Whakan slowly knelt to his knees, whilst maintaining eye contact with the deer. Thankfully, this sacred animal finally returned his slow blink and as Whakan methodically sprawled lazily across the grass, the deer relaxed and continued to happily graze. Whakan rolled onto his stomach, propping his chin up with both hands, while he and the deer had a 'blink fest', together. As part of his metaphysical training, Whakan knew the power of thought and sent several friendly thought notions the deer's way.

Whakan lay in the tall grass for some time. While admiring this delightful creature, he silently developed a plan to gain her trust and receive a gentle touch from her. He clutched a handful of grass to lure her closer to him; gradually creeping forward, when they were both suddenly engulfed in a large patch of shade, as if a huge dark cloud had blotted out the sun. Looking skyward, he witnessed an unforgettable sight. A huge, chevron-shaped object was moving slowly east, approximately 100 feet above where he lay. Rolling onto his back, he was able to see just how huge this other-worldly object actually was. Its full size was beyond belief.

Whakan thought back on how long ago his last 'trip' on Peyote was, but discounted this as the reason for viewing such a unique manifestation as what lay above him. With this thought in mind, he turned toward the sacred deer, yet she had disappeared. He smiled to himself, thinking that the 'little minx' was somehow connected to this giant object in the sky that he had just witnessed.

Bounding to his feet, Whakan knew what he had to do. He would follow this object as far as he could and make a mental note of everything he observed, so his tribe's elders may properly decipher what Whakan's thoughts and reasoning concerning the witnessing of this event could possibly mean. Quickly gathering his gear, he purposefully left the sharp, gray arrow point in the grass, as a token gift to acknowledge the exact spot where he had sighted this sacred albino deer and unknown object as it continued to hover above. Daylight once again emerged, as the huge airborne craft slowly moved eastward in the direction of Whakan's favorite mountain range, where he collected the best flint rock used for knapping.

Rapidly traversing the trail and following the retreating object, Whakan continued to watch this strange craft. He reflected on numerous stories told of the 'Sky Brothers' and how they had helped his ancestors in so many ways. According to legend, they had promised to one day return and aid his people in times of difficulty. He had personally seen this same 'V' shaped object painted and incised into numerous rock paintings in a distant area east of his village, near the desert drop off, yet here it was, drifting above him. Whakan smiled as he jogged carefully up a winding path that led to a ridge and overlooked his best-loved mountain. Again he smiled, whilst wondering what the Universe had in mind for him this time.

As Whakan gained ground on the craft, it actually appeared to slow down. He stopped for a breather and confirmed to himself that the spacecraft had indeed also stopped, stilling itself right above what he considered as his mountain. It was enormous and Whakan became convinced it could only be the same mystical 'Sky Brothers' that tribal elders often talked about. What was its purpose and why here, above his mountain and in this happy little valley? Whakan almost resented its appearance, knowing it could seriously interfere with his scheduled rendezvous that night with Johona. He laughed out loud, recognizing that his personal needs were now on hold, in light of this important occurrence.

Whakan took note of the five, large, circular lights that brightly emerged underneath the spacecraft; two under each wing with one large, pulsating red light at the center of this boomerang shaped craft. He could barely believe his eyes at what he was witnessing, let alone that this was actually happening. Immediately, Whakan thought of his training. If it had not been for the extensive knowledge he had received, no doubt he would wildly have begun to run in the opposite direction. Although this was still a consideration, as this unknown craft seemed to purposely position itself opposite his mountain, he began to get a positive feeling concerning their proposed intentions.

This huge mountain escarpment was well loved by everyone in the region for its majestic beauty, especially as the warm, pastel colors of the setting sun painted its towering west face. It was a tough climb to the top of the nearest ridge where Whakan hoped to still see what he had subconsciously begun to think was an alien craft of some description. He was not to be disappointed, as he finally crested the steep hill.

With this unknown craft still floating in front of 'his' mountain; Whakan slowed to a stop and quietly began to gather his composure. He realized the path he was on would undoubtedly lead to a confrontation with these 'Sky Beings' who operated this craft. Whakan immediately recalled the code of his tribe, 'bravery and generosity', and would strive to be courageous, regardless of what was about to occur. He reminded himself that he must utilize the four moods of stalking, as they had been taught to him by his tribe's spiritual mentor, being cunningness, ruthlessness, sweetness, and patience.

Whakan quickened his pace and before long, the silent, imposing structure that he had been chasing, was levitating immediately above him. At this same moment, he heard a humming sound and watched as a dozen or so acorn-shaped objects suddenly dropped from underneath the center of what now appeared to be a 'mother ship', before speeding off toward the mountain. To Whakan, it looked as though the huge craft had just given birth, similar to that of catfish in a favored fishing stream. Standing directly beneath this imposing craft, he could barely make out the symbols appearing across both wings of this unknown craft. Whakan was quickly becoming convinced that these visitors were aliens, having come from another world.

Without any warning, Whakan found himself cocooned in a brilliant blue light, as three of the acorn-shaped craft had circled back and were now 'spotlighting' upon this 'intruder'. These three craft surrounded him, with a small shaft of white light emerging from underneath each craft, striking the ground at his very feet. He could hear a humming sound and instinctively, the young Indian brave reached into one of the beams of light with his two hands outstretched; one hand showing all five fingers and the other displaying only two, giving a total count of 7. He quickly rearranged his fingers, so that one hand showed four fingers, while the other showed three, which once again added up to 7.

Long ago, Whakan had been taught by his elders of powers associated with the sacred number 7 and how the Universe had interwoven it into all, with 'all that is' being easily found when one knows where to look. Whakan was hoping these 'star beings' would understand and accept his signal, which had been taught to him as signifying physical and spiritual perfection.

The shaft of light suddenly extinguished and the next thing Whakan knew was that his whole body was being ripped apart, molecule by molecule, as he was slowly lifted off the ground upward, by a blinding beam of light.

"This is what you wanted?" he half accused himself, whilst fighting back the fear that his life was about to end. However, a calming and unknown voice then spoke to him from inside his own mind.

"Do not be afraid, as we will not harm you." With this, Whakan relaxed and thought out loud, "I'm all yours"!

2

The 'Sky Brothers'

As Whakan transitioned from having his body reconstructed aboard this huge alien craft, he continued to feel unsettled, similar to when he had initially been disassembled on the way up. Although not knowing it on this occasion, Whakan would use this transporter technique many times in the future, as well as becoming quite used to its use. For now, he only concentrated on his meta-physical training and began to practice his previously learned art of stalking. Regardless of the environment he now found himself in, he was determined to remain bold and confident, having already indulged and experienced a separate reality during Peyote cacti ceremonies. This technique always calmed his fears in troubling situations, as it made him feel he had turned himself into a 'stalking puma'.

With his eyes having finally adjusted to the dimly lit room he had been put into, he was surprised to see standing directly opposite himself, two rows of beings with large heads and even larger eyes. In the front row, four were much smaller than the other four standing behind. Whakan immediately felt that their large almond shaped eyes could see right through him and into his very soul. It was at this moment, he began to feel comforted as he considered himself as a 'friend to the Universe' and mentally reflected of how this was already known to these alien beings that now stood before him.

A thin smile escaped Whakan's lips when he saw these aliens were all holding hands. Two of the smaller ones approached him, reaching forward to take his right hand. Without a word said, they led him from the large chamber where they currently stood, down a smooth and shiny corridor and into another even larger room. Whakan noticed a hive of activity, as this room was filled with creatures of all shapes, sizes, and descriptions. It appeared to him as being a market of some sort or other and even in his wildest of dreams, he would never have imagined such a site.

A gentle tug on his big hand quickly brought him back to his senses and he continued to follow his new 'friends' until they entered yet another hallway that emerged into an even larger chamber, which was dominated by a large window situated at the front of this spacecraft. Within this window, Whakan's favorite mountain was framed and clearly visible from anywhere in the room.

To see his majestic escarpment from this height, and in the middle of the afternoon, brought a proud smile to his face. Whakan approached the large window alone, but then noticed numerous blue lights darting back and forth at the base of the mountain. He began contemplating the motives of these strangers. As if reading his mind, a low rumbling voice came from behind him and said, "They're relocating the animals".

Whakan had already presumed he was in some sort of command room. He turned slowly to his left, to address the person whose deep voice had just spoken to him. Two tall beings, with long white hair and fair skin, stood on either side of a large man like creature, who was comfortably seated in a large, ornate chair. One of these white-haired beings was obviously female and her beauty actually startled Whakan. The large seated male was clearly the one in charge. He had a seriously stern face, as if having survived many battles and was one who would not suffer fools easily. Whakan instantly knew that he had to 'up his game'.

The rugged appearance of this large off-worlder was imposing, but strangely enough, not in any intimidating way. Whakan sensed a fundamental goodness about this creature and instantly felt that this was someone he could readily relate to. This stranger's garments appeared custom tailored and dark, shiny material accentuated his muscular frame. His face appeared handsomely symmetrical and he definitely 'smacked a bit' of an animal, of some kind or other. He had the bearing of a well-traveled warrior, with his large mustache and earring segmented into his left ear completing such a look.

"Welcome aboard our ship, the *Soho*. This is our leader, Captain Alom, and I am known as Tugor" said the male alien, who stood directly behind his Captain. Whakan was instantly surprised, as his native tongue was being spoken very clearly by Tugor. Alom stood up from his chair and stepped forward, and Whakan was further surprised that he was not as tall as his own self.

"Greetings Alom," Whakan said warmly, as he firmly grabbed Alom's outstretched wrist, who responded in kind. "I am surprised to hear my native language spoken so well, especially when it would appear that you and your friends are clearly not from around here," Whakan continued.

"I have traveled to many lands," Alom responded in a low and husky voice. "I am much older than you can imagine and have learned many languages. I knew your ancestors well and am here to keep a promise to them."

"Our Elders often recall such promises through songs and stories, as well as in the rock etchings made by our ancestors depicting this very ship. They said the 'Sky Brothers' would return again one day and carry us to safety. Have you returned to take us away and if so, what plans do you have for this mountain?" Whakan skeptically asked.

On catching his breath, Whakan began thinking how strange it was to be conversing with these exotic strangers who had arrived from the heavens. He became determined to make the best of this situation, although he also intended to learn all he could of their customs, so he could proudly represent his tribe. Thankfully, Whakan thought, he wasn't 'high' on Peyote or anything else. He was a warrior with a thirst for knowledge and was determined to behave as such.

"I am aware of your fondness for this mighty mountain young Whakan," Alom said with a chuckle.

"It has a natural beauty, which has also captured my own attention. When the setting sun paints the peak with such vibrant colors, it reminds me of my home far away within another Universe, along with my loved ones who patiently await my return. I have brought you aboard the *Soho*, Whakan, to consult with you concerning an idea I have for your mountain."

"This is as much your mountain Captain Alom, as it is mine," Whakan responded. "I cannot own the mountain, air or land. I can only enjoy them and give thanks for their abundance. So, what is it that you have in mind?"

"I have decided to leave a sign on 'our' mountain to remind you of our friendship and the important tasks that lay ahead for you, whenever you see it," Alom solemnly alleged toward Whakan.

"What kind of sign?" Whakan replied, feigning mistrust.

"I will reconstruct this mountain to resemble my own handsome face," Alom responded straight-faced, although he paused for effect, as the beautiful female alien and First Officer Tugar had begun to laugh out loud.

"In my travels, Whakan, I have left many such signs on countless other planets. These signs have been made for the inhabitants and their descendants to ponder and dream, and to understand they are not alone in their Universe. You are yourself destined to re-enter your world, hundreds of years from now, emerging as a leader who helps your people. You will be re-born in this very valley and after discovering this image on 'our' mountain; your memory will be sparked to recall having met me, and your inner intuition will become illuminated toward your path ahead."

Whakan dropped his gaze and slowly shook his head, side from side. "Sounds like a tall fish story to me," Whakan replied, as Alom grinned. "How can you predict an event so far into the future Captain Alom?"

"Good question, Whakan, for which I have an answer," retorted Alom. "Valerie; explain this to our young protégé."

The voluptuous female alien stepped forward and around Alom's command chair, striking a pose of authority as she shifted her shapely hips to one side. With a shake of her long, silver-white hair, she tilted her head to one side and rested her right hand on Alom's massive shoulder, whilst placing her other hand on her right hip. Alom and Tugar smiled, as they noted the startled look on Whakan's face. Valerie was the epitome of womanhood, but worse yet, she knew it and wasn't afraid to use it to her advantage.

Valerie then asked Whakan, with an essence of disdain in her voice. "Can you tell through your eyes, Whakan, that you have entered another reality?"

Speechlessly, he nodded his head in affirmation. Valerie continued, but with a stern look aimed directly toward him.

"It would make sense that what we are telling you is closer to certainty, than some wild, fish story," Valerie answered. "Our mission here is deadly serious and you need to be clear about this. It is an honor, and a privilege, to be chosen for such an important role in the survival of your species. NOW, are you ready to pay attention and learn, or should we find another candidate?"

Whakan felt the heat of Valerie's wrath toward his casual indifference and he was about to respond sheepishly when Alom interrupted.

"When the spirit created the flesh in the form of this Universe, and many others, it did so for a purpose. To put this simply, it is important that everyone has the chance to play God, to see if they can make the right decisions. If they do, then they get to go back home and join the 'Godhead'. Failing such a mission, an entity can take countless life spans to learn what the soul needs to learn. Our job is to make such opportunities available, by seeding civilizations wherever we can and nurturing them toward the path of salvation."

In a less hostile tone, Valerie continued. "It was decided in the beginning to create a record of everything. Upon a civilization reaching maturity, it would then have access to undeniable facts of things as they happened in the past, as well as the present and the probable future; all of which is based on current events, as well as thoughts."

"So, where is this record of the past, Valerie?" Whakan enquired. "I would enjoy seeing my ancestors, whom clearly I do not know."

"These are known as the 'Akashic Records' and the only way to see them is through the spirit, a world you have only recently begun to explore," Valerie responded. "Soon enough Whakan, if your motive is pure, you will be allowed to enter this sacred place. It is here that Alom can see your future self as having a huge impact on this world of yours. Therefore, we have decided to help fate along by teaching you some important new realities, which your soul will find useful for your future reincarnation."

Again, Whakan felt his head swimming with so much information, which subconsciously he knew to be true. These off-worlders seemed to be honest and sincere, but the idea of his future self being responsible for saving a civilization was troubling to him; a responsibility that he did not want. What he did want was to join Johona, who undoubtedly would shortly be waiting for him at their secret spot by the creek, wondering where he was.

"I don't understand. Why me?" Whakan enquired.

"This is your soul's mission that you accepted long before entering this lifetime, Whakan," Alom responded. "Your previous lives have contributed to everything you have become and are, as of right now. You should know that some of us are more spiritually advanced space travelers who choose to stay behind to help struggling civilizations reach their goals. Your destiny is to help your future peoples qualify for inter-planetary contact and trade, which can drastically improve the fortunes of your race."

"How so?" Whakan questioned.

"One of the many great truths you will learn is that we can communicate with the atoms and molecules forming your environment. In the vast and infinite scope of consciousness, all is possible," Alom recited as if reading from a scroll. "Such knowledge can cure all diseases and extend natural life for hundreds of years."

"Interesting, but how do I go about completing this mission, which I do not remember signing up for, my wise friend?" Whakan asked.

Alom again addressed Whakan, exclaiming "Hundreds of years from now, the discoveries you make will lead your civilization to a long partnership between interstellar communities of like-minded entities, from every corner of this Universe. Once your people are 'plugged in' to this galactic community, your home planet will reflect the true spiritual nature of our existence and in a single generation, you shall reap the benefits."

"Sounds like a waking dream, Alom. How many worlds have achieved this lofty goal, if I may ask?"

"You can ask me anything you want and at anytime you want to," Alom said. "I want you to consider me as your mentor, Whakan, as we can learn much from each other. Countless worlds enjoy the freedom of the stars, but not all of them have the best motives in mind, yet they all answer to the same, highest authority, as there are rules that everyone needs to obey."

Whakan suddenly realized he was parched from chasing Lom's spacecraft earlier and needed badly to rehydrate. He also needed to gather himself, having begun to feel a bit 'stretched out' from all this excitement. Adding to his discomfort was the certainty he would be missing his much-anticipated rendezvous with his beautiful Johona that evening. Just at that moment, he noticed a playful grin on the face of the female alien, as she leaned over and whispered something into Alom's ear.

"Forgive me my poor manners, Whakan," Alom declared. You must be tired and in need of some refreshment. Valerie; would you take our young guest to the refectory, returning within the hour, as the sun is almost in its proper position?"

"Whakan," Valerie purred. "If you would like, I will lead you to some refreshments."

"I would like that very much, Valerie," Whakan replied, with a smile. As he turned to follow her, Alom rose from his Captain's chair and grabbed Whakan by the shoulder.

"Don't be gone long; we're almost ready," Alom added.

"Ready for what Alom?," Whakan asked.

"We'll discuss this upon your return," said Alom, with a sly look having spread across his face.

Whakan returned Alom's glare with a smile and walked away, as Valerie had already turned the corner and was heading toward another door. Whakan hurried to catch up to her. As he followed, he couldn't help but notice Valerie's exquisite figure, from top to bottom. He felt his heart start to beat wildly and wondered what the relationship was between Tugar and Valerie.

As Whakan and Valerie left the control room, Alom and Tugar turned toward each other, with a knowing look of recognition.

"Do you think we should have told him that Valerie can read his thoughts?" Tugar asked Alom.

"No! Better to let him find out on his own ... after Valerie has had her way with him," Alom grinned.

"I could use a thousand more just like him Tugar. If nature takes its course and we have Whakan's DNA to help with the constructing of our future teachers, what a wonderful opportunity the Universe will have given us. His courage and boldness will serve us well, now and into the future. This isn't the first occasion your sister has helped the 'cause' by producing quality in-betweens, but I bet she's going to enjoy this one, more so than any other."

They roared with laughter, as they left the room together.

3

'Valerie'

Valerie was about to disappear around a corner, as Whakan called out her name.

"Valerie," he said, in a quiet, but commanding tone.

She liked the tenor of his voice and slowly looked over her left shoulder toward Whakan. Turning to face him, and with a delightful smile on her face, she coyly replied, "You talking to me, Whakan?"

Witnessing Valerie's flirtatious behavior, Whakan realized this memorable adventure was becoming even more memorable, and sooner than he had anticipated. He was still young, yet he had been tested and well groomed by his older lover, Johona. His attraction to Valerie was unmistakable, but he wanted her to respect him and his commitment to 'all that is'. His mission was more important to him than his strong desire to crush Valerie to his chest and show her some of the pleasurable techniques men of this world knew. Whakan inwardly laughed, as he realized that Valerie had been practicing the four moods of stalking on him, at the very same instance that he had been stalking her. He would enjoy a good 'cat and mouse' game and intended to hold out for as long as he possibly could against any of her surreptitious advances.

Valerie stood waiting by a door that appeared to open into another area. She waved her hand toward it and beckoned Whakan to follow her. As he did, the door closed behind him and he suddenly felt a movement. Startled, he looked up at Valerie, who sweetly assured Whakan that all was well.

"We are taking this elevator to my room, where I want to show you something," Valerie said, quite matter-of-factly.

"You know, Valerie," Whakan said coyly, "I was not properly introduced to Tugar, who obviously comes from your world. Are you two a couple? If you are, then we should not be heading toward your room!"

Valerie paused and shifted her full hips to one side. With a slight tilt of her beatific face, she raised one eyebrow and said, "Why, should I be worried about taking you to my room? Are you going to attack me in there, like an animal?"

"No! This elevator should do just fine," Whakan replied, with a mischievous grin beaming across his face.

"You're pretty confident, aren't you?" Valerie purred, as the door suddenly opened.

"If you knew what I know, you'd be feeling confident too," Whakan whispered, close to her ear.

Valerie blocked the door and turned her back on him. "Tugar is my brother," she smiled seductively and sauntered out of the elevator toward her room.

Soon, they were walking through a portal that opened out to a colorful and comfortable room. The view was outstanding, with Whakan's mountain clearly visible.

Valerie walked ahead, toward what looked like an open kitchen, and punched a sequence of buttons into a panel. Whakan witnessed a blue beam shining downward, across the counter top. He was amazed to see something beginning to form from the very center of this blue light, which to him appeared to appear out of thin air.

Immediately upon this strange light turning itself off, a container appeared with something materializing within. Valerie placed this unknown substance into a dish and proceeded to grab a container that held some form of liquid. Setting it down in front of Whakan, he was astounded to see it was indeed his favorite meal; cooked rabbit and acorn cakes.

"This is what I wanted to show you," Valerie said. "The same way I just made your meal, is how Lom plans to reshape your mountain. We have learned how to communicate with the smallest particles of matter and convince them to reshape themselves into our desired form. Just as Commander Lom had said earlier, even atoms and molecules, as small as they may be, carry their burden of consciousness. We can communicate with them. This may be the most important lesson you will ever learn aboard the *Soho*, on this occasion Whakan".

"This brilliant blue light Valerie. It must be instrumental in the process of restructuring matter. I have noticed it every-where," replied Whakan.

"Yes, Whakan, it is," Valerie replied. "This is the visible light spectrum of the hydrogen molecule that is a building block of this Universe. Lom and his ship come from a Universe based on a different molecular structure, being silicon, so when he enters this dimension; he must wear the camouflage of his carbon dimension."

Whakan dug deep into the delicious rabbit stew, created out of thin air, and was very satisfied with the taste. He drew a deep drink from whatever liquid it was Valerie had given him and again, was extremely gratified. Upon devouring the acorn biscuit, which was a touch too bitter for his taste, Whakan began feeling refreshed and invigorated.

"Valerie! Does Lom really intend to engrave his face across this majestic mountain?" Whakan asked.

"He does indeed, Whakan. The location is perfect; the purpose and motive are sound and it has nothing to do with Lom's ego, as he is very humble and selfless, most of the time," she said, smiling seductively.

As Valerie admired the beautiful mountain peak that appeared just outside her window, she hastily remembered Lom's schedule for her return to his control room. Suddenly, she telepathically heard and felt Whakan admiring her well-rounded backside. Valerie sauntered toward a nearby couch, leaning forward to adjust some cushions, giving Whakan a much better angle for him to admire her shapely assets.

"The main idea, Whakan," Valerie continued, whilst maintaining a serious demeanor, "Is that in the future, you will be the first to recognize this giant face on a local mountain. It will trigger the release of important inner knowledge, leading you to unravel some great mysteries, which will help convince your civilization that space travelers are not only real, they are also peaceful. We have been nurturing humankind since the beginning. Your future discoveries, Whakan, will only be possible because of all the work being done behind the scenes, to make such opportunities possible".

Whakan turned and walked toward her, joining her on the couch whilst admiring the panoramic view, in all its glory, of the entire mountain.

"What would happen, Valerie, if in the future I do not recognize the signs and fail to fulfill my mission?" he asked.

"Sadly, that is a real possibility," Valerie replied. "There are no guarantees. All we can do is set the stage. Freewill is the rule of thumb here. However, from your earliest memories, there will be signs presented along the way. Clues that hopefully, and eventually, convince you to believe and act upon your beliefs."

"Instead of Lom's rugged face on the mountain. Valerie, we should instead engrave your beautiful face. Now, that would make me remember, for sure," Whakan smirked.

Valerie smiled and thought that she would give this young earthling something to remember alright, as she seductively stretched out on the couch while admiring the lean figure and handsome face of this bold adventurer.

"Well, you know Whakan, one of the best ways I know to create powerful memories is during sex," Valerie said, with a dry, serious look on her face.

"Funny, that's what I heard too," Whakan replied, feigning disinterest, while he slowly ran the back of his hand up and down the length of her well-formed leg.

"I'm serious, Whakan," Valerie protested, as she gently began caressing the bare skin of his muscular forearm. She continued, saying "A strong thought form sent to coordination points comes back as a material manifestation. Isn't the best time to send a strong thought form during sex, Whakan?" she whispered.

"I don't know. I'm still a virgin Valerie. Maybe you can teach me?" he replied, with a mischievous grin, as he bent down and gently bit Valerie's right knee cap.

Valerie squealed with delight and after playfully slapping one of Whakan's muscular shoulders, she slowly unzipped her flight suit to reveal what she knew Whakan had wanted to see. He was not disappointed and Valerie involuntarily sighed, when she saw in his mind what delights awaited her. She was beginning to think that she could really fall for this lovable human hunk.

4

The Mountain

Lom and Tugar were working hard on the final draft of the three images they were wishing for Whakan to choose from as one of them, would be permanently engraved on the west facing side of this magnificent mountain for everyone aboard the *Soho* to enjoy and admire. At that point, Valerie and Whakan returned to the command room.

"Ah, just in time," Lom said sarcastically. "I trust you are fully refreshed, so we may get down to business, Whakan?"

"Actually, I feel completely revitalized Captain Lom. Thank you for your generous hospitality," Whakan answered, whilst glancing toward Valerie. "Valerie was very patient while teaching me how to send strong thought-forms and communicate with atoms and molecules," he said.

Lom glanced at Tugar and shared a private smile as Whakan approached them. "So, Lom, let's see what you are going to carve into our mountain."

"I have made three sketches for you to choose from Whakan. Whichever one you select will be the one we etch into this magnificent mountain," Lom retorted.

Whakan thoughtfully looked over these sketches, but it did not take him long to make his choice.

"This profile is perfect, Lom. Solemn, yet rugged," Whakan stated, with a hint of a grin. "However, to make sure I recognize your handsome face in the future, you should frame it by placing a complete circle around it."

Lom nodded at Tugar, who made some adjustments on a console. Soon after, Whakan indeed had his suggested circle appearing around Lom's rugged image.

Whakan laughed. "That's perfect! Let's do this one," he added, pointing to the image he had favored.

Lom stood up and walked over to a podium type structure located in the middle of the floor. After touching some colored buttons on a screen, he addressed everyone aboard the *Soho*.

"This is Captain Lom," he said. "The moment we have been waiting for has arrived. For those of you aboard the *Soho* who would enjoy witnessing this majestic mountain being transformed into a permanent image of your Captain, now is the moment to turn your attention eastward, as we convince the molecules to join us in a celebration of creation."

Lom moved his hand across the control panel and all at once, the lights in the immense control room dimmed. From the corners of the room, rhythmic music began to play. It had an earthy tone and was filled with the hum of cricket sounds, and other forest dwelling insects. The music started out low, but grew in strength through the addition of mystical sounds and drumbeats, which Whakan could not have known were from Indigenous Aboriginal Australians, whom Lom had admired and recorded during one of his earlier missions 'Down Under'.

Lom found the pounding drums and chanting voices irresistible. He instinctively began to move in harmony with the music. Valerie leaped up and joined him, clapping her hands as she did.

She danced around the room, moving rhythmically to the melody, whilst stomping her feet as a sign of respect to the cultural heritage of these native Australians.

Before long, Whakan joined in with the other fifty or so undulating figures who were now spread all around the room. He saw the joy in their faces and realized how much like his own tribe these alien strangers were; a heartfelt people, who loved each other and rejoiced together in nature and spirit.

Suddenly, the drums built up to a crescendo, although they stopped abruptly as a blue light winked over Lom, who began to decree, "Our ancestors and we have traveled to these worlds since time began. We follow the mandate of the Universe, which is to follow our hearts, for this is where higher wisdom emanates."

All across the ship, applause and shouting could be heard. Lom continued, declaring "With the transformation of this mountain, we leave a bold reminder to future generations that freedom lies in being brave and reaching for honesty, which becomes evident and everlasting."

Loud cheering erupted from the crew, just as a blistering bolt of white light shot out from somewhere beneath the *Soho*, striking the mountain with an audible crash. A brilliant blue light appeared, spreading itself from its initial contact position above the mountain and extending across all its corners, to form an image of the outline Lom had designed.

For a moment, there was complete silence and Whakan thought the transformation had not happened. Unexpectedly, the mountain began to move and the sound of grinding rocks echoed throughout the valley, as a fine dust arose from the floor of the mountain itself.

Slowly, the afternoon's strong breeze blew and cleared away the dust to reveal an astonishing sight. The mountain had indeed reshaped itself to form the rugged man/ape face of Captain Lom, with a perfect circle having been positioned around the image. Whakan, who initially had his private doubts that such a transformation was possible, became visibly impressed. He walked over to Lom and placed his hand on the Commander's shoulder.

"As difficult as it is for me to admit this to you," Whakan whispered, "Our mountain has become even more majestic than it previously was and for the rest of my life, whenever I visit this site, I will fondly recall this magical day and shall strive to remember all I have learned from Valerie and others."

Whakan turned toward the mountain and was visibly emotional, as well as having become very impressed.

"Although my future self might have difficulty recognizing who this face on the mountain belongs to," he said, "it is truly a grand improvement on the original."

Valerie and Tugar broke into laughter, while Lom gave a roar and embraced Whakan from behind with a huge bear hug; lifting up his 200-pound frame easily with a strength that surprised the tall, muscular young Indian brave. Setting him back down, Lom studied the gigantic profile etched into the mountain and silently wondered if the image was too obvious and whether or not he should further disguise it.

"It's a masterpiece, Captain," Valerie recounted, "and will clearly be recognizable in the late afternoon sun."

"I concur, Commander," Tugar added. "Although we are now burning daylight Sir and have several more stops yet to make on this journey."

Lom turned toward Whakan. "Walk with me," he said. All four left the control room together and headed for the transporter chamber.

"Our mission here aboard the *Soho* is a serious one Whakan," Lom said solemnly. "Our travels have taken us to numerous worlds, as we search for ways to help civilizations make progress toward spiritual freedom. Your world will only achieve this, once it has established interplanetary contact and trade. Courageous individuals, such as you, will make the difference, which is what the Universe demands. We are not allowed to do it for you. We can only provide signposts to help guide you along your way."

"That's one impressive signpost you just created Lom," Whakan retorted. "You know, I have learned so much in the very short time we have spent together, but it is no longer enough. You have ignited my thirst for knowledge. How is it that you can now simply abandon me?" Whakan asked.

Lom laughed loudly. "That's the spirit. Your hunger for knowledge will be quenched in time to come. We shall shortly return and continue your lessons. There is much we would like to share with you Whakan, such as a voyage on the *Soho*".

"What's your definition of 'shortly'?" Whakan enquired.

Lom turned to Valerie and Tugar, raising his chin toward them, whilst silently and telepathically requesting their input.

"We have numerous obligations on the other side of your planet Whakan that will take several months to complete. In the interim, there is much for you to do," Valerie stated.

"Metaphysical training is an important skill you will need to learn, so you can reach your altered state of consciousness. Your tribal Brujo teachings are important, as they will form a solid foundation for your continued education aboard the *Soho* and beyond, Whakan."

"Look for the albino deer, my young friend," Lom told Whakan. "When you see her, you will know that we will be close at hand." With that said, they all arrived at the transportation chamber and Lom grabbed Whakan's forearm in a farewell gesture.

"Show your people the mountain, Whakan," Lom said, in a serious tone. "Explain our position and be the leader that you are destined to become."

Lom and Tugar turned and walked away, while Valerie entered the chamber with Whakan at her side.

"There has got to be a better way to leave the *Soho* than this," Whakan grunted, as he stepped onto the platform. Valerie smiled, telling Whakan, "You'll get used to it, eventually! Wait until the whole ship gets whipped through a wormhole in space. Now, that's an experience!"

"I'm going to miss you, Valerie," Whakan said, with a mischievous smile as he tenderly grabbed her around the waist. "I still had a few more things to teach you," he added.

Whakan nibbled on Valerie's neck and felt a short 'buzz'. He was then astonished to realize that they were already on the ground. Valerie reached into her vest and pulled out a colorful metal bracelet that bore alien symbols.

"You may get your chance soon enough, earthling," she dismissively whispered to Whakan. "Lom looks forward to seeing you again, although perhaps not as much as do I," she said smiling, with a seductive tilt of her head. "This bracelet is a souvenir of your visit and a reminder of our mission."

Valerie suddenly took on a serious tone. "Our new friendship aside Whakan; if Lom decides you are not the one, we will not be back. Trust me! We will be watching your progress. Look for us, when you see the sacred albino deer".

She tenderly kissed him, turned and walked away, then stopping to turn back and face him again. With her sweet smile beaming, she boldly flashed her full, left breast. Her laughter was the last thing he heard, as she vanished upward in a flash of light.

Valerie knew she had fallen for Whakan, as his naughty, but nice demeanor was really just a cover for an intensive and truth-seeking young man. However, she had seen in his mind's eye that he was in love with another. This strange emotion of 'jealousy' was new to her and she embraced it like a delicious sip of her favorite nectar juice.

Whakan stood, as if in a trance, and watched the *Soho* move slowly eastward over the mountain, which now had a permanent reminder that 'Sky Brothers' truly existed. This massive mountain profile of Lom, who Whakan now proudly considered a friend and mentor, was destined to become a gathering place for tribes, from all over the region.

With the sun setting fast in the west, Whakan suddenly realized that in spite of everything, he still had time to make it to his rendezvous with Johona, who hopefully would be waiting for him by the stream. What a story he had for her, with some careful editing required, of course. His relationship with Valerie aboard the *Soho* was his tribe's contribution to the interstellar gene pool; nothing more, and nothing less. His heart ached for Johona and he dared not ruin his relationship with her by being overly forthright concerning Valerie.

5

Friendships Cemented

The crew of the *Soho* returned on numerous occasions and they always met with Whakan at the same spot, and on the same hillside, where they had first 'selected' him. Their reunions were joyful occasions, with cherished memories always shared between them. He would often recall his favorite experiences to all aboard the *Soho*, such as sacred ceremonies held around the campfire on the hillside directly across from the 'Old Man of the Mountain', as it had become known among his people. Ceremonial dancing amongst his tribe always commenced, as Lom's gigantic profile turned lavender, in the sun's waning rays.

With Whakan's training continuing, he learned of many things that initially had been beyond his wildest dreams. He even ventured on a two-week adventure aboard the *Soho*, having finally experienced transportation through a wormhole. Ultimately, this resulted in his emerging into an entirely different part of the Universe, referred by Lom as 'Zeta Reticuli', which is known today as a binary star system that astonishingly is located over 39 light years away from planet Earth!

In this small part of the sky, Whakan set foot on three inhabited planets. All were connected in harmonious collaboration and settled with beings of every size and description that gleefully bartered and traded their goods with endless enthusiasm. Some of these extraterrestrials traded goods, which were beyond explanation, and Whakan could barely offer a guess as to what their purpose truly was, let alone how to actually describe them. These planets had all achieved a real paradise, where no one ever went hungry, became sick, or held any ill will toward one another. Their civilizations were united in a common cause, as they had eventually become accepting of one another, having joined into a brotherhood of interstellar planets. This was the dream held by Lom for Earth and he remained hopeful of instilling such visionary thinking directly into Whakan.

Whakan's greatest experience attained from his encounters with Lom and all who had traveled aboard the *Soho*, as he always recounted at annual tribal gatherings held over his many years of life ahead, were his numerous visits to the cosmic library's Akashic Records. Lom and Valerie had taught Whakan how to still his mind, through meditation, and to connect with vibrations that unlocked his own consciousness within his body. With such awareness, Whakan found he was able to accompany Lom and Valerie on their travels to this magical hall in astral form, where he could view the past, present, and future events.

For Whakan, the glory and joy of seeing ancestors he had never met walk the Earth, was tempered by a frightening scene of the entire Earth rapidly rotating and destroying everything on its surface. Although Lom had explained this was a necessary annual event needed to maintain the health of this planet, he also told of future destruction being purposely delayed through the extraordinary efforts of Lom and other off-worlders. However, Whakan could not help but become plagued to think of how much mankind still needed to blossom and reach spiritual maturity to avoid such future catastrophic events.

This was the mission of the *Soho* crew, as it had been for their ancestors. The Universe had a need and everyone was being called upon to fulfill this need. Lom explained to Whakan that his future re-emergence into human society had been strategically planned, with the Universe playing an important and overt role to prevent such tragedy. In some future point in time, civilization would be in turmoil and it would be for Whakan to lead people to places they otherwise may not be willing to tread. His training now would serve his later self well, with inner strength and knowledge gained from his understanding that sincerity and honesty are out there, and would set his civilization free.

Over the next three years, Whakan honed his spiritual skills, as had been taught to him by Lom during their many arduous training sessions spent together. It was during this time that Valerie gave birth to their son, who she named 'Howakan'. Eventually, Johona was introduced to Valerie and young Howakan and although Valerie was gracious, and understanding, although she privately demanded Whakan give her a son as well, which in time, he did.

Eventually, the day came for the *Soho* to depart. Whakan would not see his celestial friends again for another 20 years, but he knew what was now expected of him and willingly stayed true to the path Lom, Valerie, and his other newfound friends had lain before him. Through this framework, his own tribe and other residents of local tribal communities would become well prepared, as other 'Sky Brothers' of every description could often be seen streaking through the skies, whilst hovering at nightfall in the distant darkness. Although now so very long ago, word had begun to spread that this area of Earth contained real humans, who respected the Universe, having chosen to live in unison with all that it held.

Following a long and fruitful life, with numerous descendants created both on Earth and interplanetary, Whakan finally exhaled his last breath. Fittingly, he was buried ceremoniously in front of many Elders, tribal members and those who have since become ancestral family relatives. He was secreted within a cave, hidden in a mountain that bore a striking resemblance to a great mentor and benefactor, who hailed from another world and was known to Whakan as 'Lom'.

6

A Sacred Artifact

Some 337 years later in 1956, Michael (Mike) was born in sunny San Diego, California. At the age of 12, he always enjoyed looking for salamanders every morning before the start of 6th Grade, having attended his Catholic practicing school in Lemon Grove known as St. John of the Cross. Salamander hunting was particularly good in the thick, lush grass within an old used car parking lot that was adjacent to the Church he attended. Mike had become fascinated with nature and animals, as a result of two cross country trips to his Mother's hometown of Marshall, Arkansas, USA.

Mike's grandfather, Harvey, had settled in the heart of the Ozark Mountains, to build his family a house, a barn, and a grocery store. Harvey constructed roads leading to nearby cattle yards, which held more barns used for storing hay that again Harvey had built. Behind the family's three bedroom home, was a large flowing stream, where young Michael loved to fish and swim. The whole experience was absolutely magical to him and imprinted an immense appreciation of fauna and flora.

One sunny morning and with only minutes to spare before school commenced, Mike lifted up an old piece of wood in search of salamanders, but found a prize of another kind; an authentic Native

American Indian arrow point! It was smooth and gray/green in color, with perfectly chiseled edges, forming a classic arrowhead. Mike had been fascinated by the Native American Indian culture ever since 3rd Grade when students all had to make costumes out of paper. His own masterfully created outfit had been copied from that of a Blackfoot Indian, yet here and right in front of him was actual evidence that Native American Indians successfully occupied this very land in the distant past.

As Mike picked up the artifact to examine it more closely, something strange happened. A flurry of images flooded his mind. Scenes of a young, sturdy Indian harvesting special rocks to splinter into arrow points, along with large communities of happy and social people, working together in peace and living in harmonious balance, who were one with nature.

Mike as a young boy when attending St. John of the Cross in Lemon Grove.

He was quickly brought back into reality by the morning school bell and with a huge smile on his face, young Michael cleaned off his prize, slipped it into his pocket, grabbed his books and lunch, and headed off to class, having been completely unaware that this was the very same arrow point Whakan had purposely left in the grass in tribute to his sighting of an albino deer.

Mike was also unaware that the energy instilled into this arrow point by Whakan had just been transferred into his own young self. The magnetic properties embedded by Whakan into this arrow point immediately realigned themselves into Mike's youthful thought patterns.

After school on that magical day when Mike had found his newly prized arrow point, he ran home and put it into a special wooden box. He would often take the sacred stone out of this box, just to hold it and imagine the former owner and his activities in a world completely different from that of his own. Two days later after attending school, he stopped by his favorite oasis that was located at the foot of a brisk running stream, surrounded by colorful foliage and filled with vigorously moving fish and crawdads. The setting reminded him of the Ozark Mountains where his Mother had grown up and where his own love of nature had begun. As Michael approached his 'personal garden of Eden', he was upset to discover that it was no longer there. Without any warning, his paradise had been bulldozed, only to make room for a Big Bear Supermarket parking lot.

As an adult, the bitter disappointment of losing his private and personal nature preserve was still a tough wake-up call for Mike and it reminded him that he simply could not take anything for granted. He was quickly learning to pay attention to detail in his daily life. The sleeper within had awakened!

7

Tragedy Strikes

Around this period, Mike's artistic talents were beginning to emerge. In Grade School he was rewarded with kisses when either Robyn or Debbie would walk across to his desk and grab his artwork, without any warning. With this kind of sweet reward system, he unconsciously strived to better his drawing skills by developing a photographic eye for detail.

Growing up in the 1950's, Mike was bombarded with science fiction movies about UFOs and aliens. This sparked the thrill of the unknown in young Michael, who watched every episode of the *Twilight Zone* and *Outer Limits*. One night, while riding his bicycle home after looking at scary pictures of monsters and aliens in books at his buddy Jeff's house, he became very frightened. Just as he reached the top of the hill, with his own house in sight, he noticed the dirt road on the right being pitch black; leading off to nowhere. Mike's imagination took hold and he thought these same scary aliens were down that very road. He was gripped by fear and raced his bike to his front yard, jumped off and stormed into the house, panting. His Mom and Dad were startled and enquired as to the reason for all his excitement. Michael thought about it and realized it was just his own ignorance causing his fear and admitted nothing to his parents, although his attention to the existence of the unknown was awakening.

Some days later, young Michael happened to notice a picture of interest in a newspaper, left sitting on the coffee table by his Dad. Within it was a photo, claiming to be that of a UFO. It looked seriously impressive. It had windows and was box-like in shape, appearing to hover in midair. A telephone pole stood nearby, adding an eerie dimension to the shape itself. Mike saved the article, as anything concerning this mysterious subject of UFOs now held his complete interest and full attention. After sighting this first UFO article, he started going through the newspaper daily. Just three days later, he came across this very same photo, although it was flipped upside down and had a caption claiming that it was the top of a car, lying in a flood, with the telephone pole appearing as a reflection in the water. Mike quickly fetched the original article, compared the two and discovered it was a perfect match. He was alarmed to think that the largest regional newspaper had the audacity to attempt this deception and it once again confirmed for Mike that from this point on, he would be deciding what was factual, as opposed to fiction.

Mike had just solved his first UFO mystery and took pride in unmasking the hoax; relishing having had the chance to use his investigative skills and he was ready to do it again. Decades later, he would get his wish, although in a most extraordinary way.

Eventually, Mike lost these first UFO articles and his prized Indian arrow point in a Bekins annual storage auction, when he fell on misfortune and hardship whilst in his twenties. In the end, it mattered not, as these epic discoveries were emblazoned in his consciousness and would magically transport his imagination, beyond the boundaries of his everyday life, whenever he chose to recollect them. What he didn't know, at this young age, was these thoughts had actually begun to shape his future.

On August 23, 1971, Mike's world caved in. At the tender age of thirteen, he watched on from a tall sand dune, as his beloved father crashed his dune buggy while racing at dusk on the beach in Cantamar, Mexico. It was when his Mother screamed a single word, 'crash', that it sunk in what those red scattering tail-lights implied. Mike doesn't remember how he got down to the scene of the accident so quickly. It was as though he flew the mile and a half to the wreck, where he then witnessed a horrific sight. The dune buggy's front end was curled up into the driver's compartment, which had critically injured his Dad who was bleeding to death internally from massive chest injuries.

Mike was helping lift his Dad, 'Jim', by the legs into the passenger vehicle that had finally made its way onto the beach, when his Dad gasped that he couldn't breathe. Mike watched on as his Dad rolled his eyes into the back of his head. Despite the promise he made to his Dad that he was "gonna be okay," Mike knew his Dad was not going to live.

A knock came on the camper door just four hours later, where he and his twin brother, Marty, had been holed up. Although Mike had been preparing himself, he was also trying to avoid hearing the worst of news, as Gordon opened the camper door and solemnly said, "It's time to grow up boys; your Dad is dead."

Mike's Dad, SSgt James Campbell Orrell with Mom 'Robbie' in happier times, along with SSgt Orrell's final resting place.

On arriving home in San Diego from Mexico, Michael, his twin brother and his Mom, found their home full of people; mourning the loss of his Dad. However, he just wanted to be alone. That evening, as he lay awake in his bed, his mind raced back over all the memories he had of his Dad; such as how he had turned their garage into living quarters for a homeless man named Herb whom he had found on his 'beat'. Mike especially remembered when his Dad had started a Boy Scout troop in the middle of Logan Heights called 'Any Boy Can', commenced with thanks to the financial backing of his friend and Pro Boxer, Archie Moore.

Attending this Boy Scout troop taught Jim's three sons (who were the only white boys in the troop), that you judge someone on the quality of their character and not by the color of their skin. Mike's Dad would often stop by their home when on duty and bring in his partner, Officer Johnson, who was as big and brawny as was Mike's father, only Officer Johnson was black and really cool, and funny.

Mike and his two brothers also learned another valuable lesson from their Dad, about the consequences of making poor decisions. Sometimes, pain is the only teacher that can bring home the message and their Dad would not spare the rod when needed, or spoil his three sons unnecessarily.

Occasionally, Jim would have to resort to using his Police belt across the boys' bare ass, when required. Should the wide belt not get the lesson accomplished, Mike's Dad would be forced to use a skinny belt that did; all the while reminding them that he was doing this because he loved them, with this concept only making sense to Mike many years after his Dad Jim's, passing.

Mike loved his Dad dearly, although he never cried once over his death; not at the scene of the accident, during the funeral that had a mile-long Police motorcade, or even decades after. From the very beginning, Mike was only interested in where his Dad went after death and he found no solace in any of the Catholic religious philosophy and teachings that he had been raised with. He wanted real answers and not some fairy tales that to him made no sense. What he did not know; however, but has since discovered, was that these very thoughts of his were being heard elsewhere.

8

Metaphysics and The 'Akashic Records'

In time to come, Mike found himself in his old neighbor-hood of La Mesa, California, where at the age of 17 he had moved into his first bachelor pad at Chevy Chase Apartments. This was the very same apartment complex where his former baby sitter had lived. Over the years, Mike had often thought of Peggy, as she was cute and curvy and had stirred a deep yearning within for female companion-ship.

Later on, he would enjoy recounting the numerous memorable events that had occurred at this first apartment; such as when he answered his doorbell late at night, only to find the cute and curvy gal who lived at the end of his hall standing nearly naked in her blue negligee, although reeking of alcohol.

It was sex school night for 'Mikie', as his more experienced partner taught the young lover a few techniques, which he proposed refining over the years.

Around this time, Michael also re-connected with his Dad's best friend Dennis, who lived just down the street from Chevy Chase Apartments. He was Mike's former baseball coach in Little League and personified the rugged image that Mike imagined he would one day become when he matured. Dennis not only took Mike on as his apprentice, teaching him how to make teeth as a dental technician, but he also revealed his passion for the study of metaphysics, which is basically the search for real answers to the Universe. Mike was 'commanded' by Dennis to read the works of three authors, in a very set and particular order, and he could expect to be grilled on what he had learned from the reading of these works the following day. This was the landmark event that would first introduce Michael to unknown and genuine accuracies that would eventually set him upon the path Captain Lom and Valerie of the *Soho* had envisioned for a re-incarnated Whakan.

The first author young Mike was to read, was named Carlos Castanada; a University of California, Los Angeles (UCLA) student undertaking his thesis on hallucinogenic plants of Native American Indians. In his search for an experienced guide, Carlos was eventually introduced to an authentic Yaqui Indian sorcerer, named Don Juan Matus, with Carlos becoming his apprentice. The adventures written about by Castaneda were thrilling and roused in Mike an insatiable thirst for more of this raw and bona fide state of being. This is where Mike first learned of the existence of a 'separate reality', which is also the title of one of Castanada's books.

Carlos would purify himself and then partake of the Peyote cactus in an elaborate ceremony, with his mentor Don Juan nearby. It was great timing to be reading this material for Michael, as he had just been introduced to marijuana, which he found expanded his consciousness and allowed his mind a place to travel.

The second writer Mike was 'ordered' to read, was that of T. Lobsang Rampa; a Tibetan monk who had been sent west by the Dali Lama to write of his experiences growing up in a Tibetan Lamasery. Rampa had been trepanned; an operation that drills a small hole through the front of the skull. According to Rampa, this enabled him to see the auras of those visiting the Dali Lama to ascertain if they were being truthful, or deceitful. This procedure is also referred to as the 'opening up the third eye', to see what it is that really exists, with Rampa's writings becoming pivotal to Mike who for the first time, was introduced to the existence of the Akashic Records. These Akashic Records are a cosmic library, or compendium, and are believed to probably have been created by aliens, containing encoded emotions, feelings, views, and happenings, all of which are referred to as the 'etheric plane'.

In Buddhism teachings, it is said that the Akashic Records contain everything past, present and probable future, with the past being critical as it affects that which is current concerning each and every individual. Through this understanding, Rampa was able to visit other worlds by going into a meditative trance, which he used to experience and describe remarkable events that could shape human history. For Mike and chief among Rampa's theories was the disclosure that not only does aliens and UFOs exist, they are emissaries from God who seeded our planet and have been nurturing mankind since our beginning.

One of Rampa's books titled *The Hermit,* revealed a fascinating tale of a blind monk who was taken aboard a UFO and hooked up to a device, allowing him to see. This blind monk was then able to debate with and partake amongst, the minds of numerous alien species.

Another of Michael's favorite tales by Rampa, was when he had been trying to make his way west after the Chinese invaded Tibet and when he was then discovered by an attack dog, whilst hiding in the city. Rampa, who growing up had been taught telepathy in the Lamasery, sent a happy thought form to this vicious dog, whose demeanor instantly changed. The dog left, but quickly returned with a gift; that of a severed human hand in its mouth, which he presented to Rampa. Unfortunately, the guard dog's strange behavior had attracted the attention of its handler, who grabbed Rampa and took him to the barracks where the guard dogs were kept. Apparently, the dogs had taken over the compound and killed and eaten several of their Chinese handlers. Rampa was told that he would be thrown in with the dogs and was to gain control of the compound; as otherwise, the Chinese would be in big trouble from the Russians, as they had been the ones who had trained all the dogs.

Rampa requested that the dog he had befriended be allowed to accompany him and as soon as they were locked in the cage together, they were attacked by other dogs. The one dog Rampa had brought in with him fought off all the other dogs, except for the alpha male who lunged for Rampa's throat. With a deft side-step and punishing thrust to the alpha male's throat, the dog lay dead, even before it hit the ground. The Chinese soldiers watched in utter amazement, as Rampa gave silent commands to all the dogs and they proceeded to move together as one.

This idea that you could send and receive thoughts without speaking sunk home within Mike. He had grown to love all animals and enjoyed silently communicating with them, just as Rampa had. Later on, Mike experienced his own telepathic series of events that cemented his belief in the existence of untapped human abilities. Slowly but steadily, he was beginning to realize there truly was a separate reality and not that which was being taught in schools. It made him sad, as well as frustrated, to realize all he was missing; simply because his civilization was spiritually ignorant.

The third piece of work he was to read appeared to have been purposefully left for last by his new mentor Dennis, as it was rich in spiritual reality. It would represent the closest thing for Mike to a new age *Bible*, in that the source of information it contained came almost directly from the spirit world, which was intense.

This author confirmed the writings of the previous two, being Castaneda and Rampa, which Mike found comforting. This confirmation cemented within Mike his spiritual foundation, as 'truth and loyalty' had already become his religion, with 'bravery and generosity' his code, as was the Native American way; and now, it was his. Again, he had awakened!

This third author's name was Jane Roberts, with her work collectively known as 'The Seth Material', with it explained that the term 'all that is' relating to God and the preservation of memory with a renewed creativity concerning one's own self.[1] Jane began as a poet, but then discovered automatic writing, which is said to be a subconscious form of writing, rather than an intentionally, conscious one and written by spiritual beings. Before long, Jane began experiencing and going into a meditative trance with her spirit guide; a personality known as Seth, who eventually took over her body and begin dictating to Jane's husband, Robert. The transformation from Jane to Seth was apparently quite remarkable and among other quirks, Seth would smoke, while Jane never did.

At this stage, Mike had no problem believing that this spirit channeling could actually happen. It was the same process that America's famous 'Sleeping Prophet', Edgar Cayce, had used in the 1930s to predict events and cure the ill, even though the patient was half way across the country. Edgar would lie down on a couch, go into a deep, mutated sleep, and his spirit guide (or someone else) would give readings to others.

Seth had direct access to this Akashic Records cosmic library and the information he revealed through Jane from this extraordinary

---

1 New Awareness Network, *Seth: The Spiritual Teacher that Launched the New Age* (online) 2003 <http://www.sethlearningcenter.org/>.

source was startling, to say the least. Mike found comfort in what was revealed about death as this was his original quest; to discover where his Dad went after his body died. Apparently, we are as dead now as we are ever going to be, with consciousness being the only true vehicle that survives immediately after the flesh sack has expired.

It matters not whether science can prove this or otherwise, as one day, the knowledge of it will be known to all. We do not have to die to meet God, which drilled home to Michael how frustrating it is that our current civilization is not harnessing their spiritual capabilities. Life could be so magical, instead of the daily grind we have actually created for ourselves.

There was one uncovering in the Seth material that shook Michael to his very core and eventually became a foundation stone that shifted his spiritual awareness to another level, as all apparently is not what it seems in life. The Christian *Bible*, as many experts would agree, holds countless contradictions that are all dependent on what story the writer was telling concerning either the same or similar events. A revealing piece of information made by Seth is in the Akashic Records, for anyone to witness themselves providing they have mastered the ability to go into a meditative trance and with a clarity of purpose, travel to the Akashic Records in their astral body.

If Tibetan monk author T. Lobsang Rampa could do this with his teachers and visit other worlds, Mike began thinking and now believes that everyone can achieve such a feat, with it being time for mankind to shift gears, and get in touch with one's inherited psychic abilities to allow for mankind's next phase to develop.

Prior to Michael being introduced to metaphysics, his earlier spiritual awakening had begun when in 7th grade and attending St. John of the Cross Catholic School in Lemon Grove, he witnessed a volunteer of the Church steal money out of the tithing basket. Until then, he had been happily lost in thought, singing hymns high up on a balcony with the Church choir and above a flock of parishioners.

Lost in his thoughts, Mike happened to look down from his position to see one of the four ushers standing right below him, reach into the donation basket, which was full of money and envelopes, grabbing a handful of notes and coin and shoving it all into one of his pockets. Mike was horrified! In one bombshell moment, all his impressions of what an adult should be was destroyed. He suddenly realized that from then on, he would have to decide for himself what was real and what just a story was told by adults to pacify children.

Mike didn't know it, but this was the exact attitude the Universe had wanted him to have! He subconsciously then found himself gazing at the large figure of a crucified Christ that stood behind the podium where the priest was speaking. Mike blocked everything else out of his mind and simply focused on the image itself.

The hypocrisy of that event forced Mike to question everything that was being thrust at him in his waking world. This incident opened the door to Mike's frustration concerning certain interpretations of *Bible* scripture. He had always been saddened by the traumatic execution of Jesus, who was innocent and not deserving of such a cruel fate. As he did, thoughts of the crucifixion event that had been bothering him for a considerable period of time again crept into his mind. Why would God allow these barbarians to torture His only Son like that? Didn't God say, "This is my beloved Son, in whom I am well pleased?"[2] "Ask, and it shall be given you, seek and ye shall find; knock, and it shall be opened unto you".[3] Jesus

2 Matthew 3:17, King James Version.

3 Matthew 7:7, King James Version.

was asking to be spared the agonizing suffering that accompanies a crucifixion, yet according to traditional *Bible* interpretation; God ignored his Son's requests and allowed him to be savagely executed.

How is it that God, who was 'well pleased' with his son Jesus, failed to answer Jesus' most fervent prayer, "Father, if thou be willing, remove this cup from me", although Jesus added, "nevertheless not my will, but thine, be done?"[4] With Jesus having asked God to take this Cup from him on three separate occasions, His Father still allowed Him to be brutalized by the barbaric civilization existing at that period. This did not make sense to young Michael, despite the fact that crucifixion was predicted in the *Old Testament* and Jesus had asked for God's will to be done. Christian dogma could well be interpreted as not only painting God as heartless, but also someone who threw Peter and Judas under the bus, making them out to be cowards and traitors.

Mike wasn't buying into any of these slanderous interpretations, even at the tender age of 13. There had to be a more logical and practical explanation for what the *Bible* text actually said about what really happened in Jesus' final days. It would take years before Michael got his answers, which happened one night with the turn of a page, in a book that changed his life forever.

4 Luke 22:42, King James Version.

9

A Divine Conspiracy

According to the Akashic Records, a Divine Conspiracy existed whereby the historic Jesus was actually not crucified and neither was Judas the betrayer of Jesus. Instead, he actually saved Jesus by kissing a self-styled Messiah, who believed that he and not the historic Jesus, was the one to fulfill Old Testament prophecy. When Peter denied Christ by saying three times "I do not know the man", it truly was because he really did not know the man.[5]

Mary, the Mother of Jesus, attended the crucifixion purely out of sympathy for the man being executed. Roman guards in attendance later claimed that the body in the tomb was removed, and it turns out they were actually genuine with what they were saying. Those who were 'in' on this conspiracy knew that the man who was being crucified was not the historic Jesus and thus removed the body of this imposter from the tomb.

Following the death of this unknown man, after three days Jesus would appear before his disciples, saying "Peace be unto you".[6] Being the great psychic that Jesus really was, He outstretched his hands to all 11 of his 12 disciples, with Apostle Thomas not having been present or by his side when this event occurred. Instead,

5 Matthew 26:72, King James Version.

6 John 20:19, King James Version.

'Doubting Thomas' asked Jesus if he could see his hands, so he could view "the print of the nails", and thus be able to stick his finger into the puncture wounds, which surely should be laid bare.[7] From then on, Michael embraced the view that eventually, Jesus' own body became an embarrassment to himself, who had then willed and spirited himself out of his own flesh.

This realization rocked Mike's world and his beliefs to the core, as it ripped apart his inherited and traditional beliefs concerning what supposedly occurred at Jesus' crucifixion. With Mike having put his trust in this new idea of a cosmic library existing somewhere in time and space, it was now apparent that he was being tested, as if to see whether he could cope with the logic and reasoning that would undoubtedly make him a heretic to his Christian faith.

The source for this explosive leak that contradicted the accepted version of the crucifixion event, is once again shown within the Akashic Records and created by some exalted beings who had the idea of crafting a permanent record of everything that has ever happened, is happening, and probably would happen in future events; such as those that Nostradamus and Edgar Cayce most likely tapped into. Michael now speculates that one day, the entire population will be familiar with such a cosmic library and use it to reshape mankind's future by re-aligning our moral character to reflect the psychic reality that underlies our physical reality.

As Mike continues with his daily reading of various newspapers, he is continually sickened as to how barbaric mankind remains, with no lessons appearing to have been learned, even following centuries of warring amongst ourselves. With knowledge and access to this Akashic Record Hall (or Library), mankind could construct a new course towards conciliation with other enlightened civilizations

7 John 20:25, King James Version.

who have already achieved interplanetary travel and trade. Through Michael's eyes, reshaping mankind's moral character is the key to Earth's future and knowing that you really don't get away with anything, should be a strong deterrent to all of their need to follow God's laws.

It became important for Michael to remain open-minded concerning the promising existence of the Akashic Records and possibilities being proposed through the writings of premier metaphysical authors such as Roberts, Rampa, and Castaneda. Mike had previously found repeating patterns of these writers' works having been discussed and he had already begun the practice of mindfulness and 'right thinking', and was seeing the dividends of such accepted wisdom. As his acceptance of Native American Religion relating to the telling of truths, displaying of loyalty, along with being brave and generous had already been embedded within his conscious self, he was compelled to hold all judgment on this new conspiracy theory, through the gathering of further data.

According to Seth, people who are caught up in the idea of crime and punishment and require such brutal atonement for the sins of man are 'deluded', and bent on violence. To Michael, it was the life and teachings of Jesus, which was the point of His life on Earth and not the crucifixion that seemed to just be a key to mark His arrival. The myth became the legend, even though Jesus was not crucified and Judas never actually betrayed him.

Seth continued to clarify that there are untold numbers of inhabited planets in many dimensions and they all have had their Jesus event, even though it may not have been the same personality such as that of Earth's Jesus. The Spirit created the flesh for a reason and to Mike, it was so everyone can learn their spiritual lessons using free will to do so.

Apparently, every inhabited planet has received a Divine emissary from God who set the example of how everyone should strive to live by with the ultimate goal of one day being qualified to return home by ascending through a series of spiritual dimensions, each one more advanced than the previous one.

The revelation of this 'Divine Conspiracy' theory actually struck a chord with Michael, who had long ago doubted the Catholic interpretation of the crucifixion event itself. As controversial as it now seemed to be, Mike reveled in the notion that for God, nothing was impossible and indeed it had been His will for the historic Jesus to have actually been spared the horrific psychic and physical trauma that victims must endure when crucified. When Jesus fell upon his face praying for God to remove his burden of being crucified, a new door was opened for Mike that allowed him to believe in an alternate ending for Jesus, which for Michael, made his heart soar like a hawk.

Mike had never liked the idea that this innocent man was allowed to be tortured by a barbaric society. Except for one or two of Mike's friends who he discussed this theory with, reactions by some toward Mike were completely different; even extreme, to the point of physical violence. They refused to believe that the historic Jesus was not physically crucified and would not accept even the possibility that Jesus did not die on the cross.

Despite Michael's best efforts to present them with all the evidence he had gathered, he realized that they simply represented millions of Christians who would not easily or readily accept a revision to this central belief concerning the redemption of mankind. The conspiracy theory for Michael proved that God, or the Godhead, is truly omnipotent and merciful, worthy of our love, and our obedience. After all, "It is the glory of God to conceal a thing, but the honor of Kings *is* to sort out a matter."[8]

Eventually, the hard evidence proving that the Divine Conspiracy had really happened found its way across time and landed at Michael's doorstep; convincing him that his thinking was based completely on fact. Mike's first thought was to apply his metaphysical beliefs, which had already revealed the support, or otherwise, of these new ideas concerning the crucifixion of Jesus. At the top of the list regarding 'unknown truths' is the disclosure, through metaphysics, that 'thoughts have substance' and that our thoughts are not merely random or whimsical imaginings to be tossed aside after their brief spark inside the human brain. Instead, thoughts are full of power and are being recorded and observed by unseen, yet authorized entities. After all, the stronger the thought, the more powerful the manifestation perceived.

Even the *Bible* confirms the importance of thoughts, stating "Ye have heard that it was said by them of old time, though shalt not commit adultery: but I say unto you, that whosoever looketh on a woman to lust after her hath committed adultery with her already in his heart", which clearly implies that our thoughts are being monitored.[9] How else would God have known that Abraham truly intended to sacrifice his son as a burnt offering, as had been commanded by God?[10] God then saw that Abraham was willing to be obedient and this was all He wanted to know before commanding the Angel to stay Abraham's hand. Where, Michael then wondered, did Jesus show to God his willingness to fulfill the ancient prophecy that the Messiah would be crucified?

8 Proverbs 25:2, King James Version.

9 Mathew 5:27-28, King James Version.

10 Genesis 22:2, King James Version.

Michael came up with his own theory. In the *Garden of Gethsemane*, right after Jesus prayed three times for the burden of enduring his own crucifixion and again asking to have this cup taken away from him, the *Bible* describes how Jesus sweated blood, although he continued to pray earnestly.[11] It is here that Michael believes the historic Christ lived through the crucifixion, demonstrating to His Father the willingness He had to fulfill Old Testament prophecy and when Jesus exuded blood from His head. He was physically living through the actual crucifixion.

Michael was attracted to the logic of the possibility and occurrence of the Divine Conspiracy as He was not constrained as others were through the teachings they had received into a rigid and traditional system of belief. Why would Peter promise Jesus that he would never betray Him and then later boldly state three times, "I do not know the man.".[12] Peter was not a coward or a liar. Jesus was a great psychic having a direct connection with 'all that is' and was/is able to tap into the same source that other, less divine humans have tapped into; being the Akashic Records.

Jesus saw the probable future that Peter would be questioned individually and on three different occasions, being a maid and two other men, with Peter having denied each accusation, but once Peter had spoken the words "I know him not" immediately, the rooster crowed as had been predicted by Jesus.[13] Peter, who may not have been aware of the conspiracy, realized after the rooster crowed that this is what Jesus was referring too and why he had wept, which was probably out of pure joy.

---

11 Mathew 26:39, King James Version; 22:44, King James Version.

12 Above n 5.

13 Luke 22:57, 22:61, King James Version.

Michael surmises that this is why, when Peter after accepting his impending martyrdom, chose to be crucified upside down, which was not only to honor the true Messiah Jesus, but to help embed the memory of this event, so that the story of Christ would survive to modern day, and Jesus would not have to be crucified twice.

Many brave humans have paid with their lives to ensure that the message of Jesus survived, for all of humanity to see the correct human representation wanted by our creator and the love He has for us, so we could better understand why we should and need to love Him back in return.

Judas, who apparently was the one Disciple that Jesus chose to reveal this Divine Conspiracy too, courageously accepted the role of the betrayer, knowing that one day his named would be cleared and his real role in saving the life of Jesus would be celebrated. You don't spend three years with Jesus watching Him with your own eyes and how He performed the many and differing miracles that did, which could only have been orchestrated by the highest authority, without being infused with the certainty that this man is truly the Son of God; worthy of your deepest devotion, which Peter and Judas had for Him.

It was no surprise to Michael when the *Gospel of Judas* was discovered and made public and revealed what metaphysics had been saying all along, which was that Judas was the hero. Jesus confirmed this when he said to Judas, "you will sacrifice the man that clothes me".[14] Michael knew instantly what these words were alluding to. Jesus was referring to the self-styled zealot who Judas kissed that would bear the physical burden of fulfilling the crucifixion, and in so doing, secure the legend of Jesus for generation after

14 Rodolphe Kasser, Marvin Meyer and Gregor Wurst (eds), *The Gospel of Judas* (The National Geographic Society, 2006).

generation so as the stories of how our sins have been forgiven could thus be told; all because Jesus was willing to die on the cross.

At this time in history, it was important for the early Christians to believe that historic Jesus was physically crucified; as otherwise, the story of Jesus may never have made it to modern day. Many martyrs made the ultimate sacrifice in the name of Jesus, so that his life and message would resonate with millions and billions of people, forever after.

10

The 'Secret' *Message of Fatima*

Further confirmation of Michael's Divine Conspiracy theory came when Pope John Paul II when he decided to publish the 'secret' *Message of Fatima*. By order of His Excellency, the Bishop of Leiria, the third part of this message was written in secret on January 3, 1944.[15] It had been written by Sister Lucia (Lucy) Marto, who said she had a vision of a Bishop dressed in a white robe, who climbed a steep mountain. At the top of the mountain was a large cross, made of rough-hewn trunks. There, at the foot of the cross and while on his knees, the Bishop was killed by bullets and arrows. Pope John Paul commented that this third revelation was linked to his own assassination attempt and survival, but in Michael's, eyes this was an attempt by the Universe to distance the historic Jesus from the crucifixion event.

Even when Jesus lived, crucifixion was known to many as a 'shameful death', with the cross being replaced with a fish that became the symbol Christians used to recognize each other in an attempt to escape deadly persecution by the Romans. It seemed quite clear to Mike that the cross, and even Christianity itself, would in the future be replaced by a religion that more accurately represented

15 Congregation for the Doctrine of the Faith, *The Message of Fatima* <http://www.vatican.va/roman_curia/congregations/cfaith/documents/rc_con_cfaith_doc_20000626_message-fatima_en.html>.

the historic Jesus and His message of love and forgiveness that Jesus had demonstrated so well.

No wonder the Catholic Church refused for decades to release Sister Lucy's third revelation. It was because they knew it endangered the power they held over billions of Christian followers. It had been tough enough for Michael to continue attending Church services at the very Church where decades early, he 'awoke' upon seeing an usher steal money from the tithing basket; however, to suffer the image of this imposter on the cross and the warped desire of millions to blindly believe that God would not answer His Son's request to have this cup taken from him, was too much.

The last straw for Mike was attending a Christmas Mass with his Mother at St. Johns Parish and watching as the Church passed the collection basket twice, in an attempt to pay the millions of dollars in damages caused by hundreds of pedophile priests. Mike would take his love and obedience for God and His earthly persona, Jesus, within him wherever he went and not through open display with the confines of a Church.

A startling modern-day confirmation of the Divine Conspiracy occurred during the filming of Mel Gibson's violent depiction of the crucifixion in his 2004 movie, *The Passion of the Christ*. Actor Jim Caviezel, who played Jesus, told an audience at The Rock Church in San Diego how he was struck by lightning while on the cross. It occurred on the last day of filming as the actor said that he "felt a foreboding heaviness" upon him before it happened.[16]

In Michaels mind, this was a demonstration by the Universe of what Mr. Gibson was depicting in his film was not acceptable and the event was seen by numerous eyewitnesses who saw the smoke

16 Jim Caviezel, 'Passion' Star: 'You Have to Take Jesus', (August 10, 2014) World Net Daily (WND), <http://www.wnd.com/2014/08/passion-star-you-have-to-take-jesus/>.

exit from Caviezel's ears after he was struck in the head by the lightning bolt. Had the Universe approved of the depiction that Mel was filming, perhaps the sky would have opened up and rays of sunlight would have bathed the actor, instead of almost killing him due to God's wrath. It was another sign of our creator's mercy that He spared Jim's life and yet to Michael, this event was a clear sign that God does not require such self-sacrifice and that anyone who demands such a payment for our sins is 'deluded'. After years of study and research, Michael was certain that the historic Jesus was not physically crucified, which proves that for our creator, all things are possible. Jesus had indeed chosen well by selecting Judas, out of all the apostles, to reveal the mysteries of the Universe too. Judas undoubtedly knew that his name would be cursed for millennia, but it was worth it to save Jesus and fulfill his mission that had been placed upon his lap by the Universe.

The anguish Judas must have endured, knowing that his true love of Jesus would be overlooked and replaced by the false mantle of a betrayer and Michael is sure that this is why Judas took his own life, so he would not be tempted to reveal this Divine Conspiracy and save his honor. To have done so would have been catastrophic since the olden day population was incapable of comprehending the deeper significance of Jesus' mission on earth and the message of His Divine presence would never have made it to modern day, if humanity had not believed that Jesus was crucified.

The myth had to become the legend and until the moment came for when this real logic could be revealed, and digested by open-minded, rational free-thinking people, conversations that took place between Jesus and his sole follower, being Judas Iscariot, will continue to be overlooked.

Michael had been quite comfortable with the mystical readings he had absorbed thus far from the writings of his chosen three authors and their discussion on metaphysics, but this 'new truth' delivered by Seth concerning Jesus as not being physically crucified, stretched the boundaries of his imagination. At first, it made him physically sick to his stomach as he contemplated the deception forced upon him through his Catholic upbringing. If he rejected this conspiracy idea outright, it would affect the credibility of everything he had learned so far in metaphysics and he wasn't prepared to abandon such wellsprings of reason. He vowed to keep an open mind concerning the proposition that a Divine Conspiracy did exist, and undertake further research to either confirm or deny that such an event happened. He dared not share even the idea of a Divine Conspiracy with others and instead, Mike searched for years to obtain further evidence. In the end, the evidence he found was just so overwhelming and supportive of a Divine Conspiracy theory that Michael happily moved the location of the crucifixion from Golgotha to Gethsethame.

In the end, it was Mike's courage in accepting these new views that opened the door to his own life-changing event on July 1, 1990, when he captured on 35mm film, the best evidence for extraterrestrials ever produced.

11

Not Just Coincidence

Metaphysics not only revealed to Mike what really happened to the historic Jesus, but detailed the survival of our souls immediately after death, which was something else the Catholic Church didn't appear to want the masses to know, for whatever reason. By now Michael had become convinced that his Dad had indeed not only survived death, but was probably still actively working from afar to aid his sons in whatever capacity he could.

In 1999, a new television program hit the airwaves called *Crossing Over*. Mike's new belief system would pay off handsomely when his own father was successfully channeled to him in front of an audience of over 500 people on January 15, 2016, by the host of this television show being famed psychic medium, John Edward.[17]

17 John Edward is an internationally renowned psychic medium who's public and private readings have born witness to a world of spiritual and extrasensory phenomena. John's ability to use his psychic abilities, on behalf of the living who are themselves of varying ages, cultures, abilities, and backgrounds, has been well documented through researched and published works such as *The New York Times*, the *Los Angeles Times*, and *The Washington Post*. John has over 25 years experience as a psychic medium, having hosted his own TV show branded *Crossing Over*, syndicated across continents such as the United Kingdom, the United States of America, and Australia. John has successfully helped thousands of people worldwide to connect and communicate with deceased relatives and friends, through his ability to converse with those who have traversed to the 'Other Side'. The uniqueness of John's work can be further explored through his JohnEdward.net website titled *Communicate. Appreciate. Validate. Evolve.*

John appeared to accurately 'read' people in the audience through the channeling of their dead relatives, who apparently are not dead at all.

Mike, appearing with John Edward on January 15, 2016, following his 'reading' received from John on channeling Mike's Dad in front of over 500 others, with his Father then confirming to Mike through John that he knew he was the inspiration behind Mike's research into the writing of this book. Brief discussions with John on the topic of UFOs also ensued.

It was amazing that such happenings actually occurred for Mike on such a personal level, as well as on national television, with John Edward confirming everything Mike understood from his own interpretation of metaphysics. John mentioned that the dead can sometimes contact us through electromagnetic means like the telephone. This is exactly what Michael experienced during a 'eureka' moment, shortly after his Dad had passed.

Mike's family had begun to receive telephone calls, although nobody spoke on the other end of the line. His Mom would scream and blow Jim's Police whistle into the receiver, begging whoever was making such calls to cease. Mike himself answered their 'phone on numerous occasions, only to hear silence. How interesting it now is for him to know that it was most likely his Dad, who in his own way was trying to make contact with his family.

Just days after Mike's eureka moment of discovery, something strange began to happen at his apartment. Someone was ringing the doorbell and running away before Mike could answer his door. He was well familiar with this prank since he and his friends had often played it together in his neighborhood when growing up. However, it was now happening to him.

On one occasion when his doorbell rang, Mike was right next to the door and opened it up almost immediately, only to find that no-one was there. He lived in the middle of a long hallway that opened into a triangular courtyard. Mike reasoned there was no possible way any human could have accomplished an escape, in such a quickened pace or occasion, with their being only one plausible explanation.

Michael's instinct told him that his deceased Dad was manipulating the electrical connection in the doorbell itself, in an effort to make personal contact. With this sudden realization in mind, Mike spoke audible words to the invisible spirit of his Dad, telling him that he knew it was him; he appreciated his efforts to contact him and that he loved him. The doorbell pranks immediately ended from that point onward, but this was not the case concerning other future mysterious incidents.

One evening, at this very same apartment when Mike had just taken his almost nightly bubble bath and was smoking his Tijuana mall cigar whilst drying off, he faced the mirror and suddenly, without any warning, the reflection of his face in the mirror slowly morphed into the exact likeness of that of his Father's, but only for a few seconds as Mike's fear of losing his own identity shut the event down.

Michael's frustration at losing control of his fear was numbing and he resolved in the future to be more of a warrior for knowledge, should such opportunity present itself again. It seemed to Mike as if the Universe was showing him that nothing was impossible.

On another occasion, Mike and two of his male friends were at the beach when he realized why his best friend, Rocky, refused to go into the ocean above his waist. Rocky had never learned to swim, which was something Mike had simply assumed he had. As staggering a revelation that this was to Mike, nothing would compare to what was about to happen. While Rocky and his other friend, also named Mike, chatted together about 20 yards north of his position, our Mike was busy catching his quota of waves by bodysurfing. After one particularly big wave, Mike did an underwater somersault to end his ride and as he did so, his right hand dug into the sandy bottom of the ocean's beach.

During this flip and upon breaking to the surface, Mike was astonished to see that one of the fingers on his right hand had speared a metal ring that held a half dozen or so keys attached to it. He yelled at his friends, who were further north, and became dumbstruck to watch the other Mike searching through his empty pockets, only to discover that whilst body surfing, Michael had actually found his friend's missing keys in the vast ocean, before anybody knew they were even missing. The implications of this feat would weigh on his mind for decades.

Another incident that happened around this same point in time also held similar inferences. Mike's buddy Bob was an amateur astronomer and would frequently invite Mike to drive to the Laguna Mountains, where they would set up Bob's telescope together. After hooking up power from the car's battery to the telescope, it could track the stars they were observing with the telescope compensating for the earth's rotation. It was truly an eye-opening experience for Michael to see these galaxies and unique star formations that suddenly appeared to be so close.

On one such outing, Bob viewed a Ring Nebula through his telescope, which is thought to be 2,000 odd light years away from earth in the northern constellation known as 'Lyra'. Bob signaled for Mike to take a look and as he did so through the eyepiece of the telescope, a smile broke out across his face, as Mike began to compare the celestial ring to that of the smoke rings his Dad used to blow at him when smoking his Lucky Strikes cigarettes.

Suddenly, while Mike was still looking through Bob's telescope, an asteroid, or meteor, pierced the exact center of the Ring Nebula and streaked across from right to left. Mike shouted out to Bob as to what he had seen, but from the underwhelming response suggested by Bob to Mike, he did not appear to be believed. It mattered naught to Michael, as the Universe had once again shown its hand to him, in a beautiful and mysterious way.

There were other occasions when only the 'hand of God' (or an assist from his team), saved Michael from certain death, such as when he drove through a rain squall out of Gila Bend, Arizona, and the small car he was driving began hydroplaning. No amount of counter steering was going to correct his vehicle's drifting off the road and becoming airborne.

Mike had been traveling for miles on an upraised portion of the desert freeway and as his life passed quickly before his eyes, he waited expectantly to be flying airborne, landing with a sickening crunch and be 'spat off' the road. He kept waiting and waiting, yet the car kept spinning on a level surface as if God had put out his hand and held the car aloft. Mike's car had found the only turn-out for miles and this miraculous appearance had saved his life. He knew, at that very moment, he was living on extended playtime and owed a debt to 'all that is'.

Of course, there was also the instance when he fell asleep at the wheel, driving from one construction site to another. Fate placed a shallow ditch in-between Mike and oncoming traffic, which woke him up. On another occasion, Mike entered the intersection on a green light that had been green for a good five seconds, along with another car on his right-hand side. Without warning, a vehicle speeded through from the left and ran the red light, just missing Mike and the driver in the other car, positioned to his right. Michael would later recall that the offending driver who almost killed him, had his head turned to the left and didn't even appear to know that he had entered the wrong intersection.

A valuable lesson was learned by Mike that day, which also paid off on many numerous future occasions, as one should always look both ways whenever entering any intersection.

The most unusual near-death experience for Mike was when he and his college football buddy, Bill, went on a mission of 'ill will'. They were stealing hanging plants from the front of peoples' homes, late at night. After 'cleaning out' Poway, they headed for La Jolla, but the Universe had seen enough! It's pretty poor when as an adult, you make such bad decisions. However, when you have truly been exposed to the truth, as Michael had through his study of meta-physics, then you are really slapping 'all that is' in the face and to-tally disregarding all teachings.

As Mike and Bill traveled west on State Route 52, heading for La Jolla in Mike's Chevy Luv pick-up truck, they suddenly realized that a set of headlights was coming toward them and heading eastward on State Route 52, which was actually on their side of the freeway. With the two cars rapidly approaching each other, Mike flashed back to a year earlier as the same thing had occurred to him when driving alone on Interstate 805, heading north to his home in Del Mar. Someone was in his fast lane and bearing down directly toward him, but in the wrong direction.

Michael wisely avoided this oncoming vehicle by quickly steering across to the right-hand side of the freeway. The driver seemed to be doing it on 'a dare' and continued to drive in the wrong lane. However, on this occasion when driving with Bill, the headlights coming toward them on State Route 52 changed lanes and again were directly in front of Mike.

As the two cars closed the distance between them, Mike moved another lane over to the right. However, the approaching car did the exact same thing. Bill sat deathly silent, as it was quite apparent that the other driver intended to commit suicide by crashing into them, head on. At the last moment, Mike pretended to change lanes to the left then cranked the wheels to the right. The other car fell for the ruse and changed lanes and both cars whooshed past one another.

Bill and Mike were left speechless as they caught up to the car that had been in front of them, which this suicidal driver had passed, in order to aim for them. This car was a Police car and was all that Mike needed to see and decide they were done with the theft of other peoples' property. Although, Bill was less convinced and wanted to continue. What Bill did not know was that Mike had just read a section of Carlos Castanada's book, '*A Separate Reality*', which detailed an ominous tale involving a pair of phantom headlights and served as a bad omen of things waiting in the offing. Mike took little convincing to believe that the Universe had stepped in to warn him that to continue on his thieving path, would end most unfortunately for him. As it was, the two thieves wound up fighting over the stolen plants. It was an invaluable lesson that Mike took to heart and never did again.

12

Creativity and Thought

Not long after Mike's metaphysical indoctrination, he and his two brothers rented a house on Eldora Street in Lemon Grove, where Mike's first telepathic event occurred. It was an event he would look back on as being solid confirmation that mankind has been fooled. How is it that we possess so many important paranormal abilities, but have not harnessed any of them as a civilization? What a different society we would have, if humanity was plugged into this spiritual internet!

It was a day like any other day when this event occurred. Michael and his two brothers were entertaining his next door neighbors, Tommy and his son Zach, as well as Tommy's curvy sister, Gina. As Mike was mindlessly washing dishes in the kitchen, his two brothers were on the couch with Gina, who was sitting in between them and soaking up the boys' attention. Tommy and his son were sitting on the piano bench and the young boy was blissfully plucking the piano keys and playing his imaginary tune.

Suddenly, and without any warning, Tommy's voice entered Mike's mind saying, "I gotta get out of here." Mike clearly heard these unspoken words in Tommy's own voice and Mike instantly knew that Tommy had to 'get out of there'.

Tom's occupation as a painter had wreaked havoc with his asthma and he frequently suffered severe coughing bouts. Apparently, he was about to have another, right in front of everyone. Mike knew he had to help his friend, so he quickly dried his hands, walked into the living room and said, "So, Tommy, don't you have to get going right about now?" Tommy rose up instantly, grabbed his son and said "You're right! Thanks for your hospitality guys. Come on Sis", and they were gone in the blink of an eye.

Tommy's quick reaction proved to Mike that he really was in physical distress and his strong emotional thought had somehow traveled to the ready receiver, who was washing dishes in the next room. It brought Mike great joy to have been able to come to his friend's aid, by acting swiftly upon receiving Tom's telepathic plea for help. It was not until Mike turned toward his brothers and saw the looks of dismay on their faces that he realized how rude his invitation for them to leave must have appeared. Mike never bothered to explain his unusual behavior to his brothers, as they would not have believed him, nor would metaphysically be 'in-tune'.

Michael's second telepathic event was as much a surprise as was his first, as experienced in Lemon Grove. He had been hiking with his girlfriend, Ronnie, in the beautiful backwoods of San Diego, when they came across a shallow stream. Just as they began to cross the stream, a young male teenager approached them from the opposite direction. Mike was mindlessly enjoying the scenery when he caught the boy's eyes through his sunglasses. As if in slow motion, the boy looked at Ronnie and then back at Mike, and once again back at Ronnie. Once again, this teenager turned toward Mike and the boy's voice suddenly entered Michael's head with these exact words ... "You're so lucky! You get to fuck your girlfriend whenever you want."

Mike quickly focused on the youth's mouth and realized that his lips had not moved at all. Once again, he thought the situation was perfect for telepathic communications to have taken place. Mike laughed, as he embraced another confirmation whereby thoughts actually do have substance, which was a bedrock realism from which his later discoveries would be born. He then smiled inwardly and philosophized on how true the boy's thoughts actually were.

Just before Michael graduated from Helix High School in La Mesa, he had created an inspiring new logo for his school that was presented to his football and track coach, who had provided Mike with a perfect role model across the four years following the death of his Dad. This logo was of a muscular Scotsman, with Mike using his own face and hair visually transplanted within the image. Mike's coach was impressed and had it mounted in the team's locker room, and later in his own office.

Two years after Mike graduated from the school, the Varsity Wrestling Team voted to have the image transferred onto their wrestling mat, as well as having it printed onto Helix High sweatshirts and more. Not long afterward, Mike's former coach led his football team to multiple California Interscholastic Federation (CIF) championships, which was a feat not previously achieved by the school. Michael had always thought, in the back of his mind, that his inspiring logo had somehow driven the team to success. Now that he knew the power of thought, Mike soon realized that he was only limited by the creativity of his own imagination.

Thoughts matter and the answer to creating our reality would be to think of what it was that was wanted and then believe it has actually happened. Thoughts facilitate an event to happen and this applies to right, as well as wrong thinking and one should be very careful as to what is thought, as well as wished for. Michael had seen this rule in action, in numerous forms, including karma instantly kicking him in the ass when he misbehaved. Steal and be stolen from! The opposite was also true. When you take care of the Universe, the Universe takes care of you and now, with the discovery of this Cosmic library of knowledge being the Akashic Records, Mike learned that everything you think, do, or say is being recorded. Nothing can be gotten away with! Our motives are also known, which is nice to know, especially when the means to a positive end requires some creative white lies, or other such nefarious deeds to achieve positive results.

13

'ORRE': Gardener of the Earth

It was during Michael's attendance at Grossmont College, following his graduation from Helix High, he made a most profound discovery. While gathering reference material for a painting he was composing toward his art major, he suddenly realized a photo he had just looked at of a popular local mountain contained something unusual. Mike picked the photo up and to his complete amazement, yet absolute delight, he saw a huge, man-ape face that was surrounded by a perfect circle; squarely placed in the middle of the mountain. It seemed as if this huge, mysterious face had been carved by some giant, eons ago.

As Mike looked for other photos of this same mountain, he reflected upon a recent discovery of a giant face on Mars, thanks to NASA's Viking 1 and 2 Explorer mission. Every human has the intimate ability to recognize a human face, but in Michael was an aspiring artist and he could recognize patterns easier than most, yet here, quite plain to see, was a gigantic face sitting on El Cajon Mountain.

Mike subsequently went about searching for numerous other photos taken of El Cajon Mountain, including those taken by him. All had this very same giant face appearing in each and every image. The real question to then be answered was how could it be that Mike had lived his entire life in this area, yet only then recognized this obvious face inside what appears to be a huge, natural circle at such a time? Curiously, when he showed one of his photos of El Cajon Mountain to his friends, hardly anyone else saw or recognized the face outlined on the surface of the mountain. They were always finding some other peculiar anomaly on the mountain that they thought was a face. Michael decided to draw a clear acetate overlay and designed the masculine face in pen and ink. When Mike next showed this image to his friends, he would lift up the overlay to the mutterings of 'ooh' or 'aah' as the masculine face clearly became visible to those who gazed upon it. It was at this point that Mike knew he was on a 'winner'.

As time progressed, Mike convinced an owner of a T-Shirt store to team up and produce some hats and shirts with Mike's handsome logo of 'ORRE: Gardener of the Earth' inscribed upon this merchandise. The nickname 'Orre' had been given to his Grandfather, on his Dad's side, who had been a Naval Air Pilot in WWI and the term 'Gardener of the Earth' came from Mike's study of metaphysics that described aliens as having seeded planet Earth, once it had cooled.

A funny thing happened not long after Mike's discovery of 'Orre'. He began having lucid dreams of UFOs. So many in fact, he kept a journal of these dreams. Most were ground observations of aerial objects, but every now and then, Mike actually found himself to be onboard one of these alien ships, even piloting the craft. He would awake to smile thinking how much he wished it was a waking dream and that mankind had indeed made contact with extraterrestrial beings, as he had no doubt that humanity would enjoy and benefit from interplanetary contact.

Michael had by now become well read on how dreams were orchestrated by unseen entities, much like a play, and that they held both disclosed and undisclosed meanings. In hindsight, Mike now considers he was actually being prepared for what was eventually waiting him in the future!

Mike had built a darkroom in the apartment he shared with his girlfriend in Pacific Beach. It was a dream come true for him, as ever since he took a black and white photography class at Helix High, he had wanted to express himself through the developing of color photography. Now, he had his Omega C760 enlarger and all the equipment for his drum processing. He became so good at it that one of his own enlargements was selected to be displayed in a prestigious photo competition, being the 'San Diego Student Art Competition and Exhibition'. Mike enjoyed this creative process and relished the moments when he would open his processing tube to see how a print had developed.

On one warm day in July of 1990, Mike decided he needed some fresh photos to take and enlarge in his darkroom. He gathered his girlfriend and his buddy Bob, and headed to Julian, to capture some beautiful vistas on his Canon EOS 650. The threesome enjoyed some friendly banter during the long drive toward their destination, stopping at Santa Ysabel, just south east of Ramona, so Mike could purchase some of his favorite fruit bars from a bakeshop called *Dudley's Famous Bakery*.

After leaving Dudley's, and no more than a few miles further on to Julian, Mike noticed a large sign near a turnout on his right-hand side saying 'Inaja Memorial Campground'. Instinctively, he pulled in and asked his companions why they had not previously seen this sign before? Mike was 33 years of age and had traveled this way on numerous occasions, yet had never previously seen, nor noticed this particular park and its signage.

Upon stopping, and whilst Mike busily grabbed his camera and gear out of the car, he heard Bobby yell at both he and Ronnie. Eventually, the two of them caught up with Bob, who had stopped walking and was standing directly in front of a memorial marker located at the entrance to the park. Bob then proceeded to point out the date on the marker to the both Mike and Ronnie. Apparently, this park had been built as a dedication to 11 men who lost their lives battling a fire at that very spot, exactly one day before Michael's birth date in 1956.

INAJA Monument:

A Tribute to the Fallen

BY REGINA ELLING

Just a few short miles shy of Julian, the Inaja Memorial Picnic Ground and National Recreation Trail are real treats for visitors who take the time to stop and explore these special sites.

The grounds are part of the Cleveland National Forest and were dedicated in 1957. That was a year after the Inaja Fire of 1956. The man-made fire burned for five days and destroyed more than 43,600 acres, numerous homes, animals and humans.

Inaja was a small unoccupied Indian Reservation where a fire brigade became trapped in a canyon. Two men reached the top and pulled seven others to safety. The remaining 11 men — two Forest Service personnel and nine from the Viejas Honor Camp — lost their lives to the flames.

"Inaja" is taken from the Kumeyaay language and means "my tears."

The picnic grounds offer expansive views of the valley. Here visitors will find a peaceful

Continued on page 53

INAJA Monument Continued from page 52

stop on their way to or from Julian.

Open year-round to hikers only, the trail is a half-mile mile loop that takes about a half-hour to walk and goes from 3,560 feet up to 4,169 feet.

Views from the trail include the Santa Ysabel Valley, an overlook which showcases the San Diego River canyon and Volcan Mountain near Julian.

Up close, visitors can look for animal tracks among the huge boulders and oaks; the area is home to deer, mountain lions, numerous songbirds and many other species.

For something more exotic but harder to see, visit: www.orreman7.com/BestUFO photoever.html. On the Web site, Michael Orrell claims to have taken the best UFO photograph ever, right off one of the Inaja trails.

The Inaja grounds can be found by following Highway 78 between Santa Ysabel and Julian; the grounds are approximately a mile from the Highway 78/79 Santa Ysabel intersection. ■

THE GUIDE TO JULIAN • ISSUE 10 – 2010 • WWW.JULIANGUIDE.COM • 760-788-8148 53

A front page and internal article depicting the Memorial Monument located at the entrance to Inaja Memorial Park.[18]

18 Regina Elling, 'INAJA Monument: A Tribute to the Fallen', *The Guide to Julian CALIFORNIA* (2010) <http://www.orreman7.com/JulianGuide.html>.

Mike read the inscription on the monument and it was only then that as a true believer in reincarnation he comprehended the enormity of how his own self may well have been one of those firefighters who had died that sad day. Mike was more than interested, although he didn't make a big deal about it to his companions. Internally; however, he was on alert, as this seemed to be an omen of some importance.

14

Hike or Bike with Mike

The park was beautiful and had numerous picnic benches and campfire rings available for those who chose to visit, but no overnight camping was permitted. As Ronnie used the well-built bathroom, Bob and Mike studied a posted metal map that contained a description of a loop trail, with the highlight being an incredible vista overlooking 18 miles of landscape across the San Diego River trail located near Mission Valley. The two companions eagerly showed Ronnie the scenic photo displayed on the sign as she exited the bathroom, with the trio then hurrying off to see this wonder for themselves.

The winding trail along the way was filled with a delightful variety of plants and trees; a treat for the eyes, as well as the nose. At one spot, Mike found an interesting trail on the left that led toward large granite outcroppings, slightly hidden by trees. The bold hikers soon stood on a majestic escarpment, overlooking the entrance to Mission Valley itself. They could see a glimpse of the fauna and no doubt further treasures awaiting them further south down the trail. After taking some well-composed photos of the outcrop, Michael hurried his friends along to link again to the loop that connected with the trail.

As the three continued walking along the path, Mike began thinking about the coincidence of dates on the Memorial marker noticed at the Park's entrance. Metaphysics had confirmed to him the importance of interpreting signs, which the 'other side' can manifest in our world to help guide us along our chosen path. Native Americans lived with the notion that the Spirit was alive in all matter and again, since Michael had embraced their religion of 'Truth and Loyalty', he found himself on high alert for anything else that may appear as being out of the ordinary. Moments later, he looked up and saw something that completely stopped him in his tracks. Ronnie and Bob complained, but he wasn't budging as, after all, this was to be *HIS* expedition.

On Mike's left was a large granite slab and on the flat face of this rock face, directly in the middle, was a clear image of an eagle with folded wings. A close inspection of the dark coloration, which at first appeared as a purposefully designed Indian petroglyph was, in fact, some kind of deep algae that had randomly formed itself into the image of an eagle. If this wasn't enough of a reason to stop, then just above and to the right of this strange eagle-like image, was a rock outcropping that clearly resembled a human head with a chiseled jaw.

Later trips undertaken over the years to this same spot provided even better images of this rock face, as the sun was in a higher position and thus created shadows, which sculpted the image nicely. Michael explained his discoveries to his partners, while quickly taking some great photos of these gifts, but his hiking buddies lacked Mike's unique sense of creativity and missed the moment.

Soon, all three were climbing a short, although somewhat steep hill that had been nicely laid with railroad ties, which unfortunately was obliterated during the devastating Cedar fires in 2003 where over 280,000 acres was destroyed due to Santa Ana winds rapidly aiding the fire to devour this picturesque landscape.

The threesome completed the climb and found themselves on a plateau that weaved its way south to a bend in the trail. As the trio rounded the corner, there it was; a magnificent valley, bathed in blue light that appeared to disappear off and into the distance.

An observation platform had been constructed at the precipice, having used the natural rock formation beautifully to overlook this scenic vista. A metal periscope device had been erected on a wall built from rocks, so visitors could view a distant mountain that also gave an indicator at the bottom of the device, allowing for the observed mountain to be labeled. Mike immediately noticed El Cajon Mountain listed for viewing and as he sighted the mountain in his camera, he reminded his companions of the giant face on the front of it that he had discerned years earlier.

As the sun was setting fast, and Mike having no idea how far it was to the parking lot, he quickly set his Canon camera firmly on the rock wall, placing his zoom lens on maximum and slowly squeezed off a single photo of this magical valley that lay before him as the sun washed across its backdrop in a brilliant haze of blue. Normally, he would have taken several shots to ensure that at least one of them was in focus, but he was more concerned for the safety of his party as there were mountain lions in this area and it was getting dark, and fast! As he began to lead them away from this pictorial treasure, he noticed a metal post just opposite the platform designating this lookout spot as being labeled 'Scenic Spot 7'.

Michael smiled inwardly at this discovery because he knew how special the number 7 was, through his study of metaphysics. It is a sacred number representing God and is blessed more times than all other things under heaven. Mike had already commenced his research concerning such matters, whilst having made major discoveries along the way, to prove exactly this.

15

The Famous Inaja UFO Photo

The exhausted trio was in good spirits during their long drive home and joyfully recalled their hiking experience together that day along the beautiful loop trail. The glorious vision of this hazy blue valley stuck in Mike's mind, as he carefully drove the winding road home. He looked forward to making enlargements of the photos he had taken, once Thrifty Drug store had professionally developed his negatives. The three friends vowed to return again together one day, to enjoy Inaja Memorial Park when time permitted.

With Mike having taken his only photo overlooking this magnificent valley, with El Cajon Mountain rising in the distance, none of them could have known that hovering above a distant hillside were ten, acorn-shaped objects; not of this world. Nine of the spaceships were on the east side of the valley, with a lone single craft on the west side, heading easterly toward the main formation.

This 'one in a billion' photograph was destined to reveal a lost alien code that would successfully link UFOs, or UAVs (Unidentified Aerial Vehicles), with countless ancient artifacts. The question Mike would continue to pose to himself and others years later, was whether or not these 'aliens' knew he was there? Had they knowingly positioned themselves for him to photograph them, due to his background in and knowledge of metaphysics, as Mike's belief system indeed included the possibility of alien life? They may well have known he would not abandon this golden opportunity that had been presented to him for the betterment of mankind.

Mike's accidentally taken photograph of ten daylight UFOs has produced an explanation to successfully linking UFOs to each other, as well as ancient artifacts and the Nazca Lines in Peru, which was a first in the history of UFOlogy. The objects were acorn-shaped and triangular in form. They bore similarities Mike labeled his photo 'The Best UFO Photo Ever Taken', due to a pattern found in one of the objects that is duplicated in three other separate UFO photographs taken by other enthusiasts from around the world.

One week later, Mike received his photos and negatives back from Thrifty's of the trio's expedition to Santa Ysabel. Immediately on looking through his photos, Mike noticed some strange 'dots' in the background of his one and only blue valley photograph. He studied this photo closely with his magnifying loop, paying particular attention to what he had begun to think were 'objects'.

The formation seemed to be stretching out toward the right side of the valley and Mike decided to take a closer look at the air-space west of the fleet of objects. He was delighted to then find a similar object, hovering about the same height as the formation itself. In the back of his mind, Mike was beginning to think that the impossible had actually happened, but he wanted another opinion.

Mike took his photo to show his buddy Jeff, although he was not surprised to hear Jeff telling him that the objects were nothing other than flying ducks. Mike pointed out that there were no visible wings, to which Jeff responded, "If you think those dots are UFOs, you need to do another bowl!"

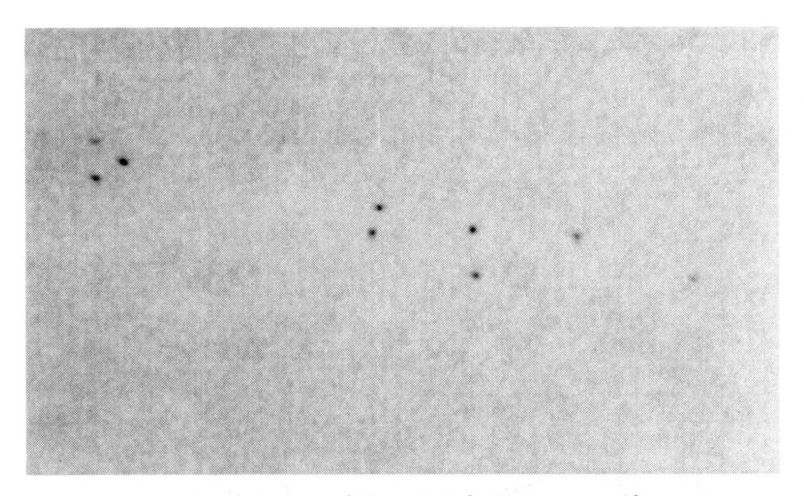

Enlargements of the UFO formation itself.

It was becoming clear to Mike that there were no wings, or ailerons on any of these objects within his now prized image, which immediately discounted Jeff's theory of them simply being birds. The discovery of these mysterious dots would not be dismissed by Mike. Unlike his good friend and fellow photographer, Jeff, or anyone else less versed in reality, he could not be swayed to believe that these 'dots' were birds, or any other such easily dismissive points of view.

Mike vowed to return to Scenic Spot 7 at Inaja Memorial Park within two weeks and film birds and planes in the same relative spot as where the UFOs had revealed themselves to him. This was exactly what he did and some two weeks later, drove all the way there for this specific purpose.

After successfully photographing birds and planes in the airspace, directly in front of the periscope platform site where his original photo had been taken, Mike turned to leave Scenic Spot 7. He just happened to look back along the trail that brought him to this point and again made a most delightful discovery. A large corner boulder was clearly shaped like an alien's head, complete with a pointed ear. Mike took numerous photos of this anomaly and wondered why it was that he hadn't noticed it two weeks prior.

The image of Mike above, caressing a mysterious alien-shaped boulder, was taken by *CBS News 8 KFMB* in 2006. Mike continues to cherish this rock formation and views it as evidence of a physical manifestation of God's creativity.

On the long drive home from Santa Ysabel, Mike was beginning to think that his personal spirit guide was secretly working on his behalf, as there had been too many unusual coincidences having taken place at this memorial park, for such thinking to be ignored. Mike's latest discovery of a large boulder in the shape of an alien head, just a dozen steps away from the exact spot where he may very well have photographed extraterrestrials, confirmed to him that he was being groomed for something 'big'. He knew the date of the firefighters' deaths, being precisely one day before his own birth, was a sure sign that fate was singling him out for some form of mission.

Mike reminded himself that you can never do enough for 'the cause' although he could not have imagined what would come, San

Diego's top television news team would send a senior producer and cameraman to follow him all the way to the park, as well as then hiking the loop trail themselves to Scenic Spot 7. Once there, they would interview Mike, as he stood on the very observation platform where he had taken what he has since called the *Famous Inaja UFO Photo.*

Photo of Mike at Scenic Spot 7.

The photos of birds and planes that Mike took at the platform site proved what he had already assumed, which was he had indeed filmed ten daylight UFOs. The question then for Mike was; what to do about it?

Mike decided to call the local experts, being the San Diego UFO Society, who told him he would need to have the objects enlarged to grain. Mike couldn't do this with his own enlarger, so he shopped around and found a photo-lab in downtown San Diego that could provide such a service.

Giant Photo Service enlarged the photo that included the fleet of nine objects on the east side of the valley, as well as the single craft appearing to hover on the west side. Once they enlarged Mike's image, they took another photo of the enlargement, in order to expand it again for accurate viewing purposes. This process would prove to be time-consuming, as well as taking weeks to accomplish.

Mike made good use of his time, whilst awaiting the develop-ment of these magnifications. With the recent purchase of a VCR, he started recording all that he could which was being played on TV concerning something, or anything, to do with aliens and UFOs. As it turned out, there was plenty to record as numerous major net-works were catering to the demand of a paranormal, thirsty public, with shows like Arthur C. Clarke's *World of Mysterious Powers, Terra X, In Search Of*, and much more.

One of the most interesting programs that he watched was an episode of *Unsolved Mysteries*, which detailed the events of December 9, 1965, when an acorn-shaped craft purposefully changed course and crashed in a forest in Kecksburg, Pennsylvania. This crash was witnessed by civilians, who were the first on the scene, and their eye-witness accounts contributed to the recreation of a life-sized, acorn-shaped craft. Eventually, the American military showed up and kicked everybody out; taking the object away, yet claiming later that it was only a meteorite that had crashed to Earth.

Weeks passed until finally, Mike got the call he had been waiting for from Giant Photo Service. He hustled downtown to collect his magnified photo of what he felt was photographic evidence of actual UFOs as depicted within these enlargements. Upon arriving home and looking at the photos, he was spellbound. The enlargement of the formation was phenomenal. Clearly, these objects were triangular in shape with no wings whatsoever, which would mean they were not of this earth, with several appearing to reflect the blue atmosphere that was below them. The implications of such a discovery were almost overwhelming for Mike.

Mike looked at the enlargement of the single craft located on the west side of the valley, but he was not impressed. At first glance, this lone object appeared to just be a dark and fuzzy blob, with no clear shape or form. He fawned for an entire week over the enlargement that depicted a fleet of UFOs, positioned over the west side of the valley. It was such an unusual feeling for him to be the first human looking at these images that for all intents and purposes, represented emissaries from God, whom metaphysics always claimed were the 'Gardeners of the Earth' and responsible for the seeding of our planet.

Michael was profoundly moved and humbled by this strange event. He decided to accept this 'burden of knowledge' and share his findings with others, whatever it was that he had discovered; yet his journey to change the world was only just beginning.

Two months prior to Mike shooting his *Famous Inaja UFO Photo*, he had completed three monochromatic paintings that reflected some of his newfound beliefs and discoveries. His metaphysical training had pointed him in some specific direction, which he willingly began to follow.

Although Mike felt he had made some historic discoveries, he had no platform to publicly share them and so he felt compelled to paint them. Most of his friends had no clue what he was talking about. When he became 'metaphysical', it was left up to him to put his hands in his head and pull something out that could be meaningful for mankind. After all, without having the courage to investigate the absurd, how can one actually accomplish the impossible?

16

Seven – The Number of Completeness and Perfection

One of Mike's metaphysical paintings titled 'Forgotten Heritage', featured several of the mysterious giant, ground markings found at Nazca Desert, which is a sequence of prehistoric geoglyphs typically formed by stones and located in southern Peru. These enigmatic figures are etched into the Peruvian desert floor in Chile and have been estimated to be over 2,500 years old. They are known to have puzzled mankind ever since they were discovered by a pilot who was flying his plane over the area, decades earlier. In particular, two figures, being monkey and duckling designs, contain unmistakable patterns that Michael now believes himself as being the first to successfully interpret, which involves the holy number 7.

It was while collecting reference material for an art project that Mike first discovered the Nazca Lines. While combing through dozens of *National Geographic* magazines at the Spring Valley Swap Meet, he came upon this global mystery that had never been mentioned in any classroom he had attended. In an issue dating from 1975, the magazine showed a picture of both the monkey and the duckling figures at Nazca, which Michael immediately noticed as containing an odd amount of fingers; 7, to be exact.

National Geographic claimed that the Nazca artist made a mistake with the fingers in the duckling figure, but they made no mention of the monkey as containing this same error. Considering the *National Geographic* is a world renowned and well-researched magazine, Mike was struck with the thought that this oversight of theirs, and even denial, was a curiosity, indeed!

With Michael having learned years earlier just how special the number 7 was and the study of metaphysics having claimed that this very number is 'blessed more than all things under heaven', after some exhaustive detective work, he discovered that this number 7 had been designed into almost every aspect of our existence. The list of 7's he collected was vast and beyond coincidence. To him, it was a clear sign or 'signature' that God indeed exists. For example, there are 7 regions in the electromagnetic spectrum, being:

1. radio waves;
2. microwaves;
3. infrared;
4. visible light;
5. ultraviolet;
6. x-rays; and
7. gamma rays.

There are 7 interstellar molecules, which are:

1. hydrogen;
2. neon;

3. iron;
4. helium;
5. oxygen;
6. carbon; and
7. silicon.

There are also 7 spectral classes of stars; 7 colors in the rainbow; 7 notes in music; 7 spinal chakras; and 7 days in a week. The *Bible* itself is a book of Seven's, such as 7 stars; 7 churches; 7 seals; 7 feasts; 7 trumpets; 7 candlesticks, and so on. Eventually, Mike was compelled to write a poem about the number 7, now depicted below:

Flight Seven

Drifting to sleep, once again I flew,
Off toward this mystery in magical Peru,
Who drew these lines on this desert plateau?
What was their purpose, he demanded to know?

I dove toward this monkey, who beckoned to me,
"Look at my hands, a secret you'll see,"
"Just like the feet on this duckling from heaven,"
"We seem to feature the holy number Seven."

"Yes, it's true," I remarked, "a pattern indeed,"
Then he rolled out a list and urged me to read,
"From the number of oceans to the classes of stars in Heaven,"
"Nothing is more blessed, than the holy number Seven."

"It represents God," he said with a grin,
"To prove He exists, He's planned it within,"
"All of your spectrums and days of the week,"
"Behold, even your Gospels of Sevens they speak."

"That explains why I'm now dying to know,"
"Please tell me wise monkey, who put on the show,"
"I would tell you young dreamer, who gets the glory,"
"But we've run out of time and reached the end of this story."

Used 735 times (54 times in the book of Revelation alone), the number 7 is the foundation of God's word. Seven is the number of completeness and perfection (both physical and spiritual). Google the *Book of Revelation* and you will find that there are 7 churches, 7 angels to the 7 churches, 7 seals, 7 trumpet plagues, 7 claps of thunder and the 7 last plagues. The first resurrection of the dead takes place at the seventh trumpet, therefore completing salvation for the Church. Add to this, the number 7 (numerated or otherwise), is sequenced over 205 times within the *Bible* (and counting ...).[19]

Michael was now convinced that the importance of the number 7 was the key to understanding the nature of reality and he vowed to continue the search for more evidence that would prove a divine force was behind the planning of this number, in all its manifestations. Indeed, he had become all the more won over than ever and with this unnatural evidence, it was clear the spirit created the flesh and not the other way around.

19 Don Ruhl, *List of Sevens in the Bible* (online), October 12, 2016, <https://sevensinthebible.com/list-of-sevens-in-the-bible/>.

17

We Are Not Alone

As these and other thoughts flashed through Mike's mind, he decided to finally examine the fuzzy enlargement of the single craft located on the west side of the valley that he had photographed that magical day in July of 1990. Already, he had recognized the peculiar fact that this single craft was hovering over the back side of the very same mountain where years earlier, he had discovered a giant natural rock face that was surrounded by a perfect circle. The periscope device at Scenic Spot 7 had pinpointed El Cajon Mountain as if it was pre-planned, here was a single UFO hovering over that exact place where the periscope's indicator lined up against this mysterious mountain.

This fuzzy blob sure didn't look like much at first, but as Michael applied his strong eye for detail and gazed at the jumble of pixels portrayed within his photograph, he suddenly realized that the left side of the object was actually in perfect focus. Grabbing his magnifying loop, he placed it on the left edge of the craft and was surprised to note a crisp and clear edge that closely resembled that of an acorn.

Mike suddenly sat up with a jolt, as he realized what it was he had just recorded on his VCR, being a television program featuring a crashed acorn-shaped spaceship in Pennsylvania decades earlier, whose contour exactly matched this very craft shown within his photograph. However, something peculiar then caught his eye. Two-thirds up from the left side of the craft was a clear spike-like projection. As he hypothesized that this obtrusion was most likely a communication device of some sort, since it was facing the 'fleet' formation on the east side of the valley, he once again realized that he had also seen this same shape in a photo taken by an astronaut that had appeared on yet another television program he had also taped on his new VCR. Mike concluded his investigation that evening, by observing the craft as having left three clear images of itself trailing the right side of the object, which accounted for the click of the shutter.

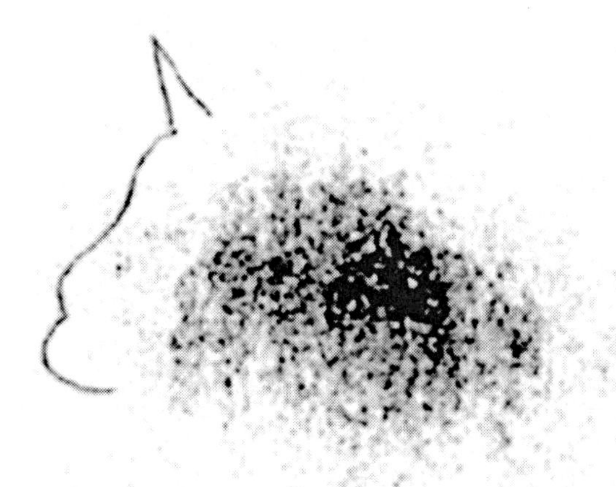

Although greatly magnified, displayed on the left-hand side of the above image (further emphasized through Mike's added drawing) is a spike-like projection, also found by Mike on numerous other UFO photos, including that of Astronaut James McDivitt.

Mike wrapped up his investigation for the night, knowing that he had stumbled upon a groundbreaking discovery and realized he had a lot of work yet to complete with the finding of these programs and the organizing all his assembled evidence. Labeling the VHS tapes that contained the corresponding evidence was just the first step.

Mike knew he would have to take photos of all these cruicial pieces of information before appearing on TV and to also use in later presentations. With a smile on his face, he thanked the Universe and went to bed. He was not surprised the very next morning to recall that he had piloted a UFO that evening, if not within his dreams.

It did not take Mike long to find the *Unsolved Mysteries* video clip of the Kecksburg UFO incident, where acorn-shaped craft matched his own photograph taken of UFOs at Inaja Memorial Campground on that hot day in July of 1990. This previous UFO event occurred on December 9, 1965, when a brilliant fireball was seen by thousands of people across six U.S. States, landing in a forest near Kecksburg, Pennsylvania.

Witnesses were recorded as seeing a "wisp of blue smoke arising from the woods."[20] The local Fire Department reported finding "an object in the shape of an acorn and about the size of a Volkswagen Beetle."[21] The headline by Greensburg Pennsylvania's *Tribune-Review*, who reported from the location, depicted the event as 'Unidentified Flying Object Falls near Kecksburg: Army Ropes off Area.'

Other accounts of this same event have been documented in numerous publications, with bystanders having stated that "the object made three sudden and deliberate changes in direction before dropping beneath the tree line two miles north of the Pennsylvania Turnpike, where hundreds of late-afternoon motorists watched it descend."[22]

---

20 Fayette County Cultural Trust, *Kecksburg UFO Incident*, <http://www.fayettetrust.org/Kecksburg-UFO-Incident.html>.

21 Ibid.

22 Clark DeLeon, *Pennsylvania Curiosities* (Rowman & Littlefield, 4th ed, 2013) 131.

Unusual markings were apparently evident and recalled as being a band of writing around the base of what appeared to be a partially buried, large, metallic, and bronze-gold in color object, with no rivets or seams displayed.[23] According to Jim Romansky, an eye-witness to this event, this object was acorn-shaped and large enough for a person to stand inside.[24]

Varying images that depict the acorn-shape of the Kecksberg UFO, as was depicted by eye-witnesses to this event.

The *Unsolved Mysteries* episode faithfully portrayed the blue light emanating from the woods where the UFO had crashed. After seeing this, Michael realized it matched the eye-witness report of his former girlfriend, Linda, who decades earlier had a frightening

---

23 Stan Gordon, 'The Kecksburg, PA UFO Crash Incident', *UFO Evidence: Scientific Study of the UFO Phenomenon and the Search for Extraterrestrial Life* (2011) <http://www.ufoevidence.org/documents/doc1300.htm>.

24 Ibid.

interlude with a UFO that emanated a strong blue light near Dehesa Road, just outside of Lakeside, California, where Mike was living.

Linda told Mike what had happened and he knew she was not one to spin a wild tale. It had apparently started months earlier, when she would occasionally spend the night at her girlfriend's house on Dehesa Road, El Cajon, California. On several occasions, they spotted this blue light hovering above a tall tree on their property. One night, Linda awoke to her girlfriend's screams and ran to her bedroom. Upon opening the door, Linda stated that she saw the entire room flooded with a blue light. Terrified, both women then ran screaming from the room.

Two weeks later, Mike found himself on the very property where Linda and her girlfriend had their strange encounter. He inquired as to where they saw the object hovering and they pointed to a large tree, about a hundred yards to the east of the house. Fancying himself as an amateur detective, Mike walked over to the distant eucalyptus tree, passing a large budding marijuana plant along the way. As he approached the tree, he looked for any landing pod marks around the base of the trunk. Not seeing anything, he looked up toward the tree for signs of broken branches. There were none. It was not until Mike turned around to walk back to the ranch house that he saw it; a perfect view of El Cajon Valley. As dusk was setting, the sparkling lights of civilization were perfectly framed by the two hillsides and would have been a sight to see from atop the tree.

It became clear to Mike that the craft the girls saw was a scout ship observing humanity. Mike and Linda parted ways not long after, but the memory of her sighting became forever etched into his mind.

Years later, Mike was attending to his booth at the 1992 UFO Conference held at the Hilton Hotel, Mission Valley in San Diego, when an elderly couple approached with a calm seriousness that actually startled him. They appeared eager to unburden themselves of information and Michael seemed the perfect man for the job. Apparently, they lived near the same area where Linda had her UFO experience and one night, they also saw a brilliant blue light illuminate across the valley, where no light should have been. It was as simple as that, but the elderly couple was clearly and deeply moved by what they had seen, and were pleased with Michael's interpretation of the event and his philosophy about the true nature of aliens and UFOs as being emissaries from God, who seeded our planet and have been nurturing mankind ever since.

Some two or three years previous to Mike's life-changing experience at the Inaja Memorial Campground, a front page story within *The Daily Californian*, (where he himself would eventually be featured), showed a drawing by one of two teenagers who had witnessed a triangular shaped object, floating in the distance. Mike noted that the object was perpendicular to the ground, as numerous objects in his *Famous Inaja UFO Photo* had shown. It seemed to Mike that these objects had been showing up on a regular basis for some time in the east of the county; a thought confirmed by ancient Indian rock paintings near Julian depicting triangular shaped objects.

These petroglyphs located off County Route S22, clearly depict UFOs and are near where Michael later found a large boulder that was shaped exactly like a human (or alien) head. In Mike's opinion, the boulder was not natural, but carved by the Native people in the area hundreds of years ago, according to the simply growing lichen plant, covering the boulder.

18

Lost Alien Patterns Explained

Continuing with his investigations, Michael found a program called *Secrets and Mysteries*, which was a taped segment he had desperately been searching for, as it contained an image of the same pattern he had just found within his own *Famous Inaja UFO Photo*. The program was hosted by an elderly gentleman who played the ghost in the TV series *The Ghost & Mrs. Muir* and in this one particular episode, the program revealed a UFO photo that Astronaut Major James McDivitt saw, filmed and photographed when it approached the Gemini IV capsule in June of 1965, as he and fellow Astronaut Ed White orbited the Earth while passing over Hawaii.

"Astronauts Edward H. White II (left) and James A. McDivitt inside the Gemini IV spacecraft wait for liftoff." [25]

25 NASA, '*Gemini IV*' (April 14, 2008) <https://www.nasa.gov/multimedia/imagegallery/image_feature_1061.html>.

Astronaut McDivitt stated that "It had a very definite shape, a white cylindrical object. It had a long arm that stuck out on the side. We had two cameras that were floating around [in the capsule], so I grabbed one and took a picture. Then I turned on the rocket control systems because I was afraid we might hit it. I called down later and told [mission control] what had happened. They went back and checked their records but were never able to identify what it could have been."[26]

This was exactly what Mike found in his own *Famous Inaja UFO Photo* when he made comparisons to that of McDivitts's UFO photo. Indeed, a spike-like projection was displayed in the same place as presented in that of his photograph. The photo taken by McDivitts was unique, but supposedly lost by NASA. Decades later, Mike's rediscovery of McDivitts's UFO photo would prove to be pivotal and fulfill the scientific burden of evidence that he had been carrying.

Another case of a UFO with a projection protruding two-thirds out of its left side was realized by Michael in an episode of *Unsolved Mysteries*, which covered a prolific amount of UFO material presented by Dorothy Izatt from Vancouver, British Columbia, with her set of circumstances first aired on TV, on December 12, 1990. As Michael carefully reviewed her footage, he made the astonishing discovery of a third projecting acorn-shape within a group of four UFOs that had 'posed' for Dorothy outside her residence.

A fourth similar shaped UFO with protrusions was revealed by famed Russian test pilot, Marina Popovich when she appeared in 1991 on the American television show *Hard Copy*. Marina is well recognized as being a world-renowned ex-military Russian test pilot, who has since become well-known for speaking out on matters concerning close encounters with various UFOs, with crash landings having been personally experienced over her Russian homeland.

---

26 Gail Barbara Stewart, *The Bermuda Triangle* (Reference Point Press, 2009) 36.

Michael had now become completely convinced that he had stumbled upon a lost alien code and wondered if Earth's past residents had seen any projecting acorn or triangular shapes for themselves. His curiosity led him to John Coles Book Shop in La Jolla, where he would make an innovative discovery on pages 9 and 27 within Marilyn Bridges' book, *Markings: Aerial Views of Sacred Landscapes*. There, perfectly etched lines onto the desert floor at Nazca, Peru, clearly depicted gigantic projecting triangles.[27] Michael was dumbfounded. He had already deduced one-half of the mystery at Nazca when realizing that two of the giant animal figures (being a monkey and ducklings) were purposefully designed to highlight the holy number 7. Now, he had found the signature of Nazca artists themselves.

To support his hypothesis, Mike would spend as much as it took to purchase the books that contained the best evidence that he could then find. On one occasion, he spent over $160.00 on a single, huge book on Indigenous Australian rock paintings, which was full of pictorial evidence that gave countless clues for interpretation against his UFO findings. On that occasion, Mike willingly suffered the wrath of his live-in girlfriend, for spending so much money on a single book. However, he knew that eventually, he would use this knowledge gained to try and deliver his research to a skeptical public, as best he could, so no expense was too great if he had the money to spend, which was a rarity.

Sometime later and through his continuing studies, Mike came across the driving force of the UFOs that created the Nazca Lines in Peru, being 'God L', a Mayan deity, which was portrayed on a classic Maya vase with a huge hummingbird on his shoulder. Directly above 'God L' was a clearly stylized projecting triangular shape, which became a remarkable discovery for Mike to have found.

---

27 Marilyn Bridges and Maria Reiche, *Markings: Aerial Views of Sacred Landscapes* (Arpeture, 1986) 9, 27.

With Michael's appetite now whetted and yearning for more, he hit the libraries and bookstores, where he found portrayals of artifacts from seemingly every ancient civilization. Rock paintings by numerous Indigenous Aboriginal Australians to that of Native American Indians, which all bore the insignia of the same distinct projecting acorn or triangular shape that linked tribes of humans together from around the world.

Ground markings, ancient sculptures, and even buildings carried the same signature that Michael had first found on his single *Famous Inaja UFO Photo* taken that fateful day at Inaja Memorial Park on July 1, 1990. Mike knew it was more than just a coincidence that he had been gathering all this knowledge, with it now becoming connected, as if all his experiences in life now seemed pre-planned in an effort to prepare him mentally and spiritually for the monumental task that had been handed to him, which was to find a way to now convince humanity that not only do aliens exist, but they have been nurturing mankind since Earth's beginning.

It was not long after Mike had collected enlargements of his *Famous Inaja UFO Photo* from Giant Photo that he discovered a very distinctive pattern, which would lead him to the unfolding of the UFO and Nazca Line mysteries. He had learned years ago, through his artistic training and a keen eye for detail, that pattern recognition is an important skill to have. When he saw this strange, spike-like projection protruding from the single craft on the west side of the valley within his *Famous Inaja UFO Photo*, he began his search to see whether this same pattern also existed within other UFO photos. It did not take long for his hypothesis to be confirmed. This spike-like pattern was most likely a communication device used by the aliens, not only to contact each other, but as a means to identify them through the ages.

It was Michael's great fortune to be the first human to stumble upon this pattern, make its connection to UFO communication methods, and recognize its importance. Every turn of the page in the countless books he collected and research completed, led to another breathtaking discovery of this same pattern, either etched on desert floors or on rock walls and Mike had become the first to understand the significance of his finding.

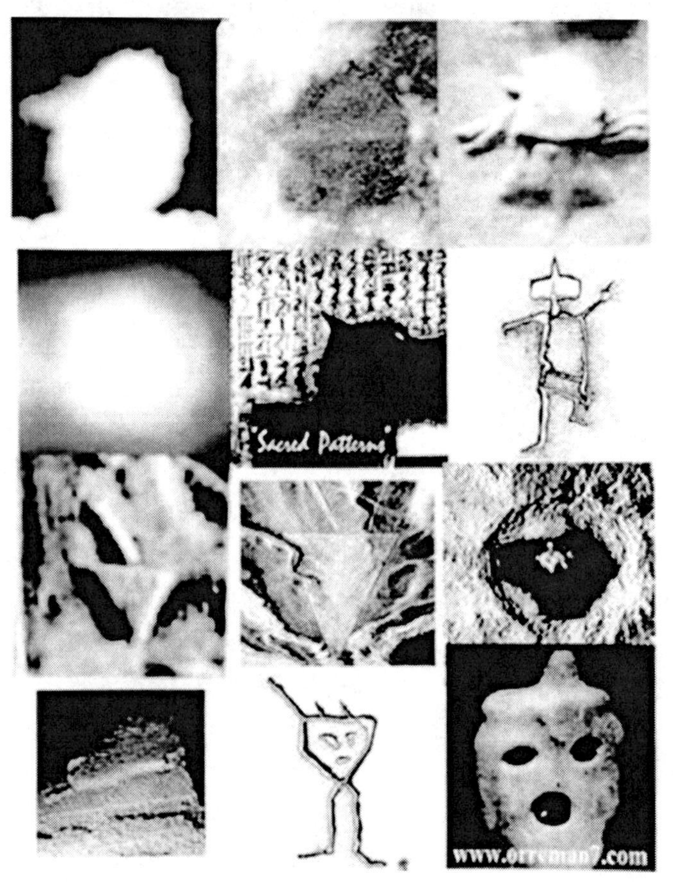

Descriptors of the above patterns observed by Mike provided below.[28]

28 Michael Orrell, *Best UFO Photo Ever 2: The Projection is the Rosetta Stone* <http://www.orreman7.com/BestUFOphotoever2.html>.

From left to right, descriptors of the image above are:

1. Photo of a UFO taken by Astronaut James McDivitt during a Project Gemini IV flight over Hawaii June 5, 1965.
2. Similar marking on the face of the moon 'Mare Serenitatis'.
3. Large carving at Petra in SW Jordon.
4. UFO photo by Canadian Dorothy Izatt with projection two-thirds up the left side.
5. Egyptian tomb ceiling with projecting acorn-shaped scored on the ceiling.
6. Ancient pictograph shows a projection of a head while pointing upward.
7. Projecting triangle in an Egyptian tomb.
8. The exact same symbol on the Nazca Plain (photo by Marilyn Bridges).
9. 150-mile wide crater on the backside of the moon is a projecting acorn-shape.
10. The Top of Cheops' Pyramid is a projecting triangle.
11. Kachina petroglyph exhibiting a walking acorn-shape with projection.
12. Mayan artifact that has contours of the Kecksberg UFO with projection.

Even prior to Mike's epic discovery of these spike-like patterns on UFOs, he had realized the importance of recognizing number patterns. No-one else had connected that the monkey and duckling figures on the desert floor at Nazca had the same pattern of 7 fingers, which Mike would later discover was embedded in every facet of the cosmos. It seemed quite apparent to him that our creator purposefully designed the number 7 into the very fabric of our existence so that mankind would have a 'map' that enabled us to eventually realize that there truly is a supreme intelligence behind the construction of our Universe and our purpose to exist.

This is exactly why aliens have purposefully tried to expose countless generations of humans to their communication devices, knowing full-well that people from across the many and varying corners, and continents of Earth, would record things over time through their interpreted rock art, geoglyphs, basketry and more. Clearly, these aliens were trusting that someday mankind would recognize these important, repeating patterns and embrace them for what they actually are; being the key to the commencement of communications leading toward mankind's journey for the initiation of interplanetary contact and trade.

According to Mike, one has only to look at the moon to see this pattern that has the ability to set mankind free. The exact same projecting acorn-shape that Mike captured on 35mm film years earlier, is still the most recognizable and unique formation on the Moon's surface, being 'Mare Serenitatis' or the 'Sea of Serenity'. This is clearly visible from Earth, without any aids or use of binoculars or telescopes needed, and it is Michael's belief that aliens purposefully formed this pattern on the Moon. Michael has been the first to recognize it for what it is, being a beacon of truth announcing that we are not alone in the Universe.

The evidence for aliens on the moon continues to stream onto the Internet, as armchair astronomers around the world capture on video shadows moving across the moon, as well as strange formations and tracks of unknown origin.

Mike was astonished one day when a book he had bought called *Celestial Raise*, which included a word for word transcript recorded by HAM radio operators when Neal Armstrong and Buzz Aldrin experienced their UFO encounter as the Eagle landed on the moon in 1969, as a part of their Apollo 11 mission. At that time, recordings obtained indicated that transmissions from Apollo 11 to NASA included the wording, "These 'Babies' are huge, Sir! Enormous! Oh my God! You wouldn't believe it! I'm telling you there are other spacecraft out there, lined up on the far side of the crater edge! They're on the Moon watching us!"[29] Some theorists claim that the moon might even be hollow, as scientific experiments showed after NASA crashed a vehicle onto the Moon's surface and recorded the ringing afterward, which should not have happened on any solid planet.[30]

There were other important patterns that Mike was first to uncover, such as the sequence of lines and their exact spacing he had first seen on the *Unsolved Mysteries* depiction of the Kecksberg UFO, as was presented using eye-witness accounts recorded within this program. Mike also noticed these same lines and their spacing were displayed at Petra, Stonehenge, and Tulum, as well as in ancient Egyptian artifacts. Whether by good luck or guidance, Michael has determined it is this very pattern that is the visible light spectrum of the hydrogen molecule, which he assumes aliens also know and use for their own space travel across this Universe.

Right next to Stonehenge is a gigantic, acorn-shaped ditch and until Michael saw it, much conjecture as to what it was and what it could be used for has been documented, although consensus seemed to concur that it remained undecipherable. This same acorn-shape is also seen in Native American Kachina drawings, as

29 Starship Earth: The Big Picture, '*NASA Forgot to Tell Us About the Transparent Dome and UFOs on the Moon in 1969*' (February 23, 2013), <http://www.starshipearththebigpicture.com/2013/02/23/nasa-forgot-to-tell-us-about-the-transparent-dome-and-ufos-on-the-moon-in-1969/>.

30 The Event Chronicle, '*6 Astonishing Irregularities That Are Evidence of a Hollow Moon Space Station*' (August 27, 2015), <http://www.theeventchronicle.com/metaphysics/galactic/6-astonishing-irregularities-that-are-evidence-of-a-hollow-moon-space-station/>.

well as other ancient artifacts, but the most unique and awe-inspiring example is a vast hole in the Universe, where no stars exist and the shape of it is also that of a magnificent acorn. Coincidence? Mike thinks not!

19

A Burdening Knowledge

As the first to find this lost and treasured information was the same for Mike as it must have been for English Egyptologist and archaeologist, Howard Carter, when he discovered the fabulous tomb of King Tutankhamun early in November 1922, along with King Tut's remarkable artifacts that were left behind. However, in Mike's case, the treasure he found has the potential to shape mankind's destiny.

It was a euphoric feeling for Michael, thinking he had been hand-picked by the Universe for this unique honor, but at the same point, he knew he had inherited a huge burden of knowledge to share and was now responsible for finding a way to do just that on behalf of his own civilization. He vowed then and there to continue his search and to find even more convincing evidence that would eventually persuade journalists, producers, and anyone else who saw his work, that mysteries surrounding UFOs had indeed been solved.

With Mike's natural curiosity and belief system intact, he was able to make some major discoveries that an average human would not think possible. He knew, through his study and understanding of metaphysics, that the number one rule was "in the vast infinite scope of consciousness, all is possible."[31]

Literally speaking, this quotation purports that atoms and molecules, minute as they may be, carry their own burden of consciousness. If one is spiritually adept, as aliens no doubt are, they can communicate and convince molecules to behave in a certain way, which is undoubtedly how the historic Jesus was able to transform three fish and two loaves of bread into enough food to feed over 4,000 people. With this fact in mind, nothing was off-limits in Michael's pursuit for the sharing of this truth and he soon made a discovery that literally was 'out of this world'.

It was not long after Mike's research led him to the finding of a clear, projecting acorn-shape incised into the wall above the entrance to the famous Treasury Building located in the ancient Jordanian city of Petra, South West Jordan, he then found the very same image embedded onto the surface of the moon. The lunar crater known as 'Mare Serenitatis', which is a huge volcanic dark patch on the surface of the moon and although visible from Earth with the naked eye, is the largest projecting acorn-shape that Mike would find, or so he thought.

Apparently, there is an immense dark space in the cosmos where no light from stars or distant galaxies is visible, which is also in the shape of a projecting acorn. The Universe seems to have gone to extraordinary lengths to identify the emissaries and their signature craft, as Michael personally believes.

One of the discoveries Michael had made left him with no doubt that he had truly stumbled upon the means of solving some mysteries concerning UFOs. Within one of his many expensive books

---

31 Peter Wilberg, *From New Age to New Gnosis: The Contemporary Significance of the New Gnostic Spirituality* (New Gnosis Publications, 2003) 77.

that he had purchased on the topic of archaeology, he found at first glance on the ceiling inside the tomb of Egyptian Pharaoh Seti I, a painting existed showing a giant projecting acorn-shape with a male figure's hand literally pointing inside of the projection itself.

This is a clear signal to any observer that this hand is pointing to something important that the artist would want to be seen. Without Mike's prior discovery of the spike-like projection in his own *Famous Inaja UFO Photo*, the riddle of Seti I's ceiling may very well have gone unnoticed and undeciphered. Mike would later discover in *Nature* magazine, a Maya sculpture that duplicates the Seti I tomb ceiling, with the right-hand side edge remaining a perfect representation of the acorn-shaped profile as depicted in Mike's own *Famous Inaja UFO Photo*.

For Mike, the real clincher that the nine objects visible in his *Famous Inaja UFO Photo* that appear to move in strict formation together through the sky were actually extraterrestrial, was because three of them on the extreme left, had fashioned themselves into a perfect triangular configuration, which would indicate these objects as being under some form of intelligent control.

Whether or not these objects were occupied by alien pilots was unknown to Mike, with considerable thought, evaluation and input from others still required. However, it was the direction that this triangular formation was pointing toward that alerted Mike to what became another obvious conclusion, which was that the entire formation had purposefully stretched itself westward and toward the approaching single craft.

The fact that these objects may very well be the same emissaries repeatedly discussed and talked about within the study of metaphysics, completely floored Mike. How was it that he had qualified to be the recipient of such an honor? He was certainly aware of how average he was, as against others, but was there something that made him unique? Initially, he thought it was his love of nature and animals that he had acquired from his Mom, as well as his compassion for humanity given to him by his Dad. After all, it was his father's death that had spurred Mike to discover the reality behind our very existence. The Universe; however, was apparently assisting in his quest to unravel 'the unknown'.

20

Presenting the Evidence

It would happen as it usually did, out of no-where, and by complete surprise. There were many instances when Mike turned the pages of his archaeological books, he was slammed with the recognition of the symbols staring back at him, such as in the *TIME-LIFE* books about ancient mysteries, when Michael clearly and instantly saw within an aerial color photo of Stonehenge, a clear acorn-shaped mound right next to these giant monoliths. To Mike, it was a clear signature of the artists who created the formation, being aliens. This same acorn-shape is just like the hoard of similar shapes from numerous sources that he had already found.

On one occasion, he was thumbing through a *National Geographic* magazine when he turned the page and was stunned to see the most remarkable manifestation of the sacred projecting acorn-shape. It was billed as the largest solar flare ever and it was accorn shape with a protrusion two-thirds up its left side, just like Astronaut James McDivitt's UFO photo. He immediately closed the magazine and was shaking a bit, thinking about the ramifications of what he had just seen. It was not until after some hours later that he dared to return to the *National Geographic* magazine and slowly, respectfully, opened it up to the image, and smiled deeply. The Universe was so cool!

In 1990, personal computers were just hitting their stride and with no success spending a lot of cash faxing his humble news releases, Mike went to the Birdie-Taylor Library in Pacific Beach and got his first e-mail address. He had managed to have some local Mutual UFO Network (MUFON) UFOlogists visit his apartment, where he boldly gave them a presentation of some of the highlights of his UFO evidence on VHS tape and in photocopies. The last thing he did was to pass around the enlargement of his *Famous Inaja UFO Photo* and asked the four men to each draw what they saw in the enlargement. Mike was curious if they would detect the clear acorn contour of the crafts left side as well as the spike-like projection, observed two-thirds up the left side.

He was disappointed when not one of the drawings faintly resembled what was actually there. It confirmed for Michael that he was blessed to see this sacred pattern, but would be severely challenged to convince professionals of its value. The drawing by the UFOlogists remains a cherished keepsake in Mike's files and a reminder that he has been singled out to continue with his work.

Confident that the Universe would come to his aid, he somehow convinced a columnist with the *Los Angeles Times* to invite him to his office so he could prove his theories in person, which he did. On Wednesday, June 12, 1991, Mike obtained his first feature story with a convincing headline, 'Hey You Think It's Easy Being Chosen to Spread the UFO Gospel', labeling the evidence Mike had presented as 'UNSETTLING'.[32]

---

32 Tony Perry, 'Hey, You Think It's Easy Being Chosen to Spread the UFO Gospel?', *Los Angeles Times*, June 12, 1991.

The article accurately reported some of Mike's discoveries and included a list of VIPs and TV shows who had refused to cover his groundbreaking discoveries up until this point. However, this positive, first article started the ball rolling and Mike was soon contacted by Copley-Wireless, agreeing to be interviewed about the *Los Angeles Times* editorial piece.

Mike later used his *Los Angeles Times* article, written by Mr. Perry, as a foundation stone to win an invitation to meet the famous test pilot from Russia, Marina Popovich, with Marina having already appeared on the television show, *Hard Copy*. This program had announced that Marina would soon be giving a presentation in Malibu, California, to discuss the numerous UFO sightings she had personally been witness to in Russia.

While Michael was reviewing his VHS tape of this episode, he instantly saw on *Hard Copy* that one of the Russian UFO photos matched exactly the Nazca marking, known as 'The Concorde', which had a huge projecting triangle, with Mike subsequently taking a photo of it directly off the television screen. He felt compelled to contact the organizers of this event in Malibu and faxed them his *Los Angeles Times* article, along with photocopies of his hypothesis relating to the Russian UFO/Nazca connection. Apparently, the host of the event was impressed with his story and thus invited Michael to attend this exclusive event.

The next week, Mike traveled to Malibu and arrived at a stylish mansion where this event would be held, enjoying some tasty appetizers and refreshments upon his arrival. After listening to Marina's speech later that evening, Mike picked his moment and with evidence in hand, approached Marina's interpreter and pitched his own UFO unearthing. She was impressed enough to gather the entire Russian delegation together, which gave Michael the opportunity to bravely brief them of his own theories, with enthusiastic dialogue commencing once he had shown his photos and explained his discoveries. Mike had the privilege of making two presentations to the Russian delegates, as noted in his front page *Beach & Bay Press* article. They were visibly impressed when he offered evidence as to both Marina's and his own UFO photos displaying duplicated markings on the desert floor at Nazca, as can be seen within Mike's website and YouTube video.

Within a week of giving his presentation to Russian representatives in Malibu, Michael was contacted by a lady who had witnessed his efforts and she was impressed enough to have Michael invited to another party to honor this esteemed visitor from Russia. On this occasion, Marina's lecture would be held at a large estate owned by a real estate mogul in Rancho Santa Fe, California.

Michael arrived punctually and came prepared to give Marina copies of his UFO photos. A wonderful event transpired before he and Marina would eventually swap their photos, which occurred during a slide projection when Marina's interpreter pointed out that the photo displayed on the screen shows a sun flare captured on film and that the flare exhibits a human face. Her announcement was met with a painful silence; then she hit the remote and moved on to the next photo. Michael was momentarily caught in a flashback as the 'flare face' photo was on the screen, with him having also captured a 'flare face' in a photo he took around sunset in La Jolla. In the photo of Mike's flare, he used a loop to see what appeared clearly at least for him to be an alien's face, complete with antennas.

"Wait", Michael spoke out loudly. "Bring that previous picture back up." He then strode up to the stage, confident that he had seen the face in the flare at the last second and as the picture was re-instated, he calmly pointed out the features that clearly looked like a human's face. The large crowd hummed with acceptance, as Marina smiled broadly.

Later that evening Michael had an audience with Marina and her small delegation, with Mike explaining to her through her interpreter, that he was the one responsible for her television interview with *Channel 10*'s reporter, Jim Wilkerson. Marina was grateful for Mike's assistance in getting her the interview and she wrote a lengthy note on the back of one of the Russian UFO photos that she then gave him. He responded by presenting her with a full 11 x 14 frame of his *Famous Inaja UFO Photo*. Unfortunately, he never had her Russian writing interpreted, before he foolishly lost this photograph.

Mike left Marina with copies of the evidence he had collected and driven back to San Diego, thinking that it was in the hands of the Universe now as to what would happen next, which only fate could decide, let alone guess.

The very next night, Mike and his girlfriend were watching Channel 10 and when field reporter Jim Wilkerson introduced Marina, she was actually holding the very same 11 x 14 UFO photo that Mike had given her the night before. She held up Mike's *Famous Inaja UFO Photo* in her hands, with his photograph thus appearing on TV within the content of this interview. Michael would replay that moment for years afterward, with a smile on his face, knowing that he had followed his instinct from the moment he saw Marina in *Hard Copy* and now, here was the fulfillment of his dream; to share a remarkable discovery with important people from other countries.

This footage was vital provenance that Michael had indeed personally met with the famous Russian and the video clip came in very handy when he approached his next news objective, being the *Beach & Bay Press* in Pacific Beach, where Michael lived. His pitch to the popular newspaper not only included his UFO discovery as seen in the San Diego edition of the *Los Angeles Times*, but a remarkable find in the scenic cliff overlooking La Jolla cove, as the previous year while hiking, he had discovered a rock formation that clearly resembled a human head. Not long after faxing his material to the newspaper's Editor-in-Chief, Mike received a phone call from the *Beach & Bay Press*, with an interview then arranged with Mary Willmont, the paper's Lead Writer. Two days later, Mary showed up at Mike's apartment with an assistant and after some casual conversation, Mike showed them his UFO evidence. Mike played the VHS footage that had been shown on Channel 10 of Marina holding up his UFO photo during her interview. Afterward, Mike gave Mary some photos of the rock face in La Jolla and she was then on her way. Days later Mike was delighted to find that he had won front page coverage in the *Beach & Bay Press* with a clear photo of the La Jolla rock face he had found and a headline that read 'Pacific Beach Resident Photographs UFOs in San Diego'.[33] The informative and lengthy article mentioned Marina holding up Mike's UFO photo as seen on TV, which was a major coup for Michael as it provided a sense of legitimacy to his cause.

33 Mary Wilmont, 'Pacific Beach Resident Photographs UFOs in San Diego', *Beach & Bay Press*, April 23, 1992.

Pacific Beach Resident Photographs UFOs In San Diego

A PB resident believes three dimensional faces found in rocks at La Jolla Cove (above), are from previous visits by space aliens.

By Mary Willmont
Beach & Bay Press

Pacific Beach resident Mike Orell, 35, knows exactly what the lyrics to the song, "It's Not Easy Being Green," mean. As a newly initiated unidentified flying object (UFO) sighter, he is discovering the amount of disbelief which exists concerning aliens and UFOs.

On July 1, 1990, Orell, an account executive for Smoothreads Custom T-Shirts, accidentally photographed what he believes are 10 UFOs hovering over the San Diego River basin near Santa Ysabel.

They appeared as 10 unexplainable black dots on his photograph. His curiosity led him to have the dots professionally enlarged for identification.

The enlargements showed the formation of dots to be triangular, acorn-like shapes with possibly shiny metallic undersides. Orell then made an exhaustive search for information about UFOs.

"My search has been both exhilarating and frustrating," Orell said. The exhilaration resulted from believing that he has become a member of a select group of UFO sighters and that there is a great deal of valuable information waiting to be presented to the people on earth.

The frustration is the rejection and disbelief he has encountered since first presenting his picture and conclusions. He has been rejected by various magazine editors, television program directors and film producers.

All responses have not been rejections, however. Orell has appeared in Tony Perry's column in the San Diego edition of the Los Angeles Times three times, and he has been inter-

cont'd on page 2

Pacific Beach Resident
from page 1

viewed by KSON and radio stations across the nation. In addition, Jim Wilkerson, of Channel 10 KGTV, aired a show last November about legitimate sightings of UFOs in Russia. Marina Popovich, the USSR president of the World Association of Women Scientists, was his guest. At the end of the show she held Orell's picture in her hand, which he had personally given to her at a reception in Los Angeles earlier in the week. Orell said she received it with excitement and immediately showed it to her colleagues. In return she gave him three Russian UFO pictures, one of which she had signed.

Because of people like Popovich, Orell vows to continue his mission to validate what he feels is "the greatest UFO photograph ever taken."

"Not only is it an in-focus daylight photo of an obvious UFO formation, but there are 10 of them (UFOs) all clearly reflecting on their metallic surfaces the surrounding terrain," he said.

Orell said UFO photos are not new and they are no longer dismissed by the government as flukes of nature. Also connected with UFOs are three dimensional faces which appear on land formations throughout the world. Orell speculates that these are mementos of aliens' visits to our universe.

Orell said he has located four such faces in San Diego County. One of them is on El Cajon Mountain, sometimes referred to as El Capitan. Orell said one of the 10 UFOs in his photograph was hovering over this mountain. Orell named the face Orvi, the nickname of his grandfather, a naval air pilot killed in World War II.

Another face, according to Orel[illegible] appears on the side of La Jolla Cove natural rock wall. He photographed [illegible] four months ago. It appears as a cle[illegible] profile when standing on the rock where the waves break. The other tw[illegible] faces are in the East County mountain[illegible] and in Inaja Park.

Orell said he believes that once h[illegible] theories are accepted, the faces wi[illegible] attract people from all over the worl[illegible] "When the world finds out about o[illegible] [San Diego's] faces, they will come [illegible] see them," Orell said.

In the meantime, he plans to conti[illegible] ue spreading the word about UFOs.

Mike's editorial piece as printed in San Diego's *Beach & Bay Press*.

Not long after the article hit the stands, phone calls started rolling into the paper from interested citizens concerning Mike's discovery. Mary decided three of those calls were important enough to forward to Michael and he followed the leads, one-by-one.

Standing in front of the La Jolla Cave rock face, which Mike believes was formed by aliens, the similarity between this and the large boulder also in the shape of an alien head that he uncovered at Scenic Spot 7 in Inaja Memorial Park when taking his *Famous Inaja UFO Photo*, is clearly more than coincidence.

It was an elderly and retired gentleman whose story brought home the chills for Michael following the publishing of his editorial piece in *Beach & Bay Press*. This soft-spoken and intelligent man calmly retold Mike of a conversation he had with a contractor who worked for him in the 1950s. At the end of work one day, the two had dinner together where the subject switched to prior occupations. The contractor revealed an amazing story to Mike that had happened over ten years earlier when he was a Military Policeman on a nearby Air Force Base. The contractor said he can only talk about the incident now, because he was sworn to secrecy for ten years after the event, with this time having since elapsed.

It was 3:00 am when alarms started blaring on the Air Force Base and as the startled MPs gathered their wits, they were instructed that an intruder was on the Base, and to meet at a particular hanger. The contractor said that when he arrived with ten other Military Police, they were astonished to find a UFO hovering outside the entrance to the hanger. They disembarked from their vehicle and surrounded the object. Unbeknownst to them, there was a captured UFO in the hanger behind them, which is undoubtedly why the large UFO was now lingering before them. After a lengthy silence, the Sergeant in charge ordered his men to open fire on the vessel, which they all did.

The contractor said their bullets had no effect and were absorbed by the craft, and then something terrifying happened. A panel slowly opened up in the craft and a green flash of light struck the Sergeant who had ordered his men to fire and the Sergeant was vaporized instantly. According to the contractor, this apparently became a very clear lesson for everyone present.

It was comforting to Mike that the UFO only disciplined the Sergeant who gave the command to fire, and spared all others. The story seemed plausible to Michael, who could understand that alien weaponry would be far advanced than would be our own.

Mike would see on television years later, just how advanced alien technology really is when it was reported that UFOs in the past had shut down a number of inter-continental ballistic missiles (ICBM) in America, and then turned some on in Russia. The knowledge that aliens could control the 'switch' that could start or stop a nuclear confrontation, was re-assuring in Michael's eyes and he personally thanked 'them' in advance for intervening, should mankind's ignorance reach such a low.

Mike was elated with his front page *Beach & Bay Press* article and knew he had to strike while the iron was hot. Along with the *Los Angeles Times* article, he approached his next objective, being *The Daily Californian* in El Cajon.

The scenic valley surrounding El Cajon Mountain, is home to the world's largest natural rock face located on the mountain itself and Michael was determined to bring it to the attention of the local residents and beyond. He had discovered it years earlier while collecting reference material for a project he was doing toward his art major at Grossmont College. The image on the mountain was over a mile in diameter and it was surrounded by a perfect circle. The sight of the giant face as you first drop into the valley while driving along Interstate 8 and heading east, was enough to jar one's senses to make you wonder whether or not this obvious human face really is natural or otherwise. Not only is the giant face perfectly placed within an enormous circle, but its position on the mountain is centered as well, as if by design.

Mike had previously named the giant face 'Orre', after his Naval Air Pilot grandfather from WWI. This masculine image on the mountain faces west and when the setting sun hits 'Orre', it changes colors from orange to purple, adding to the dynamic appearance that it already expounds. Then, as if his magnificent image wasn't enough, adding to the mystery is the sobering fact that 'Orre' is linked to Mike's UFO theories that very well may change the reality of humanity.

Mike was feeling confident as he walked into the office of *The Daily Californian* and made his proposal to Billie Sutherland, who was the top female writer with the newspaper at that junction. She seemed open-minded and receptive to Mike's presentation. He was surprised; however, when the newspaper purposefully picked a Friday to run Mike's discovery of the World's Largest Natural Rock Face. The fact that he won front page coverage showed how important the papers' staff felt Mike's discovery actually was, as the paper was not published on weekends, which meant his story would run all weekend long; a fortuitous event indeed.

Is he or isn't he?

Making a Martian out of a mountain?

Mike's article as it appeared in *The Daily Californian* on May 3-4, 1992.

The headline read 'Making a Martian out of a Mountain' and featured two of Mike's photos of 'Orre'; one with an acetate overlay that contained his ink drawing outlining the giant face, and the other photo of the mountain itself.[34] The comparison presented

34 Billie Sutherland, 'Making a Martian out of a Mountain', *The Daily Californian*, (May, 3-4, 1992).

side-by-side was a great idea of Billie's and was equally fascinating to look at. It wasn't a lengthy article and Billie got some of it wrong, but the reaction from the community was extraordinary. Billie told me of how hundreds of phone calls came in from excited readers who had finally discovered that they had been living under the gaze of a giant, stern face on El Cajon Mountain. Half of the callers had no idea there was a face on this mountain and the other half claimed that 'everybody knows about the old man in the mountain'.

So much positive feedback received concerning Mike's article that weekend prompted Billie to run a follow-up article in *The Daily Californian* several days later. On this occasion, the newspaper would feature Michael's remarkable discovery of a boulder, shaped like an alien head that was positioned right next to the very spot where he had taken his *Famous Inaja UFO Photo*.

Thar's faces in them there hills

Mike's *The Daily Californian* article of May 17, 1992.[35]

35 Billie Sutherland, 'Thar's Faces in Them There Hills, *The Daily Californian*, (May, 17, 1992).

The finding of this sacred boulder shaped like an alien's head and located right next to the spot where Mike shot his *Famous Inaja UFO Photo*, proved to him that the Universe was putting in a lot of effort to orchestrate events for Michael to locate and interpret. Each occasion he visited Scenic Spot 7, Mike would pause to hug the mysterious boulder and thank the Universe for the gifts he had been given. Now, with his latest article published featuring this alien head boulder, others would hopefully take the trip to Inaja and see this one of a kind location for themselves.

Mike, with the alien head boulder he found located only 10 feet behind the Scenic Spot 7, where he by chance took his *Famous INAJA UFO Photo* on July 1, 1990.

The proximity of this unique boulder located just steps away from the platform where Mike shot unidentified aerial objects, had to be more than a coincidence, just as his *Famous Inaja UFO Photo* was itself. However, comparisons would have to wait for another occasion, as Michael was already in search of his next opportunity that would help him circulate his findings. Following the acquisition of a personal computer, he set about building a website that would help him to substantiate and deliver all his collected evidence and research so that one day he could present it to the world in a comprehensive fashion.

He couldn't have imagined on the day he had taken this photo where fate would lead him, with his work also featured for an entire year at the San Diego Air and Space Museum's 'The Science of Aliens' exhibit, which was concerned with proof about the possibility of aliens and UFOs.

Michael Orrell in front of his 'The Science of Aliens' exhibit held at San Diego's Air and Space Museum in Balboa Park.

Early 2008, Mike learned of famed UFO researcher, Paola Harris, as giving a presentation in Carlsbad, California. As such, he contacted Paola about his own UFO story. She respectfully invited Mike to attend her lecture that was held on January 20, 2008 and allowed him to speak to over the 100 people present relating to his UFO research.

Paola was impressed with Mike's account and during his talk, had projected behind him a huge image of his *Famous Inaja UFO Photo*. It was during Mike's speech that he held up an 8 x 10 enlargement of the UFO fleet shown within his photo for the accumulated crowd to gain a better look. Later, Paola posted a photo of Mike giving his lecture to this crowd and nine years on, the link to this event is still available for all to see.

Mike providing testimony of his UFO discoveries before those present who attended Paola Harris's lecture in Carlsbad, 2008.[36]

Metaphysics told of how aliens seeded our planet, which would, of course, include humans. The question remains as to why?

36 Paola Harris, Exopolitics, *San Diego Sightings are equally important!* (2007), <http://www.paolaharris.com/sandiegostory.htm>.

However, Mike came to a simple answer, in due time. Aliens, the most advanced ones, truly are emissaries from God and in an effort to fulfill the Creator's agenda, they have generated a working vessel of flesh so that our souls can complete their spiritual training and move on from this pious plane to a higher level and intensity. Like throwing a stone into a pond and watching the ripples, not knowing where his actions would wind up … maybe on fertile ground or maybe not, but his job as he saw it, was to look for every opportunity the Universe provided and make an organized and serious effort to spread the story, which was that aliens and UFOs exist, and have been nurturing mankind since the beginning.

It was frustrating to Michael that despite the solid evidence he had collected, he was still given 'the cold shoulder' by organizations and television programs that were specifically formed to explore UFOs and the unknown. Mike didn't take it personally because he knew that not only was the populace still wary of the entire UFO subject and philosophy behind it, but he was just an ordinary citizen, with no academic accomplishments, other than a Degree in Graphic Design.

Mike genuinely felt that he had just been handed the key to unraveling some of the Earth's greatest mysteries concerning UFOs and the developing of an understanding as to how they converse. How could he dare to make such a bold and brash proclamation as this?

On balance, he was at best an amateur archaeologist with an open mind, but he has always taken his obligations seriously and has succeeded in accepting this gift of being allowed to 'see' how the pieces of this puzzle would fit. He felt he had stumbled upon a real treasure that was beyond value and if properly developed, and promoted, there may even be a part to play in achieving interplanetary contact and trade; an accomplishment that could save and change humanity forever.

Mike was determined to find a way to convince others of the benevolent nature of our space brothers, who have been so poorly portrayed in our cinemas.

21

I 'AM' Music

Every day, Mike would scan through newspapers, magazines and television guides, looking for any leads that might result in the publishing of another article, or gaining of an interview. One day, two years following Mike's article depicted in San Diego's East County *The Daily Californian* newspaper, he was reading an article in San Diego's *SLAMM: San Diego's Lifestyle and Music Magazine*. The article, by publisher Donavan Roach, was concerned with the rock star Prince, who Mike had determined years earlier, was a suitable candidate for fulfilling the metaphysical prophecy of star children coming to Earth to spread love, which is one of life's greatest forces and not that of power and money, as many believe.

This was an opinion that Mike had kept close to his heart, knowing that without having a proper background in metaphysical law, no-one would understand his theories. It had already been predicted that the Biblical man known as *Paul of Tarsus*, who set up and organized the Christian religion, would one day return to correct the error he had made. Apparently, Paul was meant to establish an environment where Man had an individual relationship with God. Instead, the message subsequently became lost and stifled in the strict and secretive religion now known as Christianity.

One day in 1984, fate intervened and Michael was given free tickets to see the premiere of *Purple Rain*, which was Prince's debut movie that eventually won Prince an Academy Award. The pass for two was given to Michael by Brian, the son of his baseball coach, who won them in a break-dancing contest. Unfortunately for Brian, the movie had incurred a 'restricted' rating and Brian was too young to see it. Out of all the players on the team, Brian chose Mike to give his tickets to. Mike thanked him sincerely and on the day of the premier, he almost did not attend. Besides the fact that he was feeling a 'little under the weather', he also was not a Prince fan, having only become barely aware of him through his catchy song *Little Red Corvette*. However, a deal is a deal and Mike was a man of his word, having promised his girlfriend, Esther, that he would take her to see it. Consequently, off to the movie theater in Grossmont Center, La Mesa, they went.

After parking his car, Mike and his date walked to the theatre entrance, which is when the Universe tapped Mike on the shoulder in the form of a very mysterious event. Nearing the movie's entrance, Mike noticed a large crowd that surrounded a small van. Suddenly, someone loudly and clearly called out his name. Without hesitation, Michael responded, "yo, I'm right here, what did I win?'. Hustling towards the van before the real person spoke up, as he knew that he had not talked to anyone or signed up for anything, and as such he was instantly aware that this was an omen and that he needed to be on alert. Arriving at the van he was handed a *Purple Rain* Album and poster and he was immediately propositioned to sell them to several rabid fans, but refused, with the notion that he would present them to Brian as a consolation prize for giving him the tickets.

Michael was surprisingly impressed by Prince's charisma in *Purple Rain*, and thought that if he had have been a rock star himself that he would have performed just like Prince; and then it happened!

While singing the song 'I would Die 4 U', Prince spoke/sang the words, 'I'm your Messiah and you're the reason why'. This was totally unexpected and suddenly Mike realized that the strange occurrence outside the movie theater, when his name was called out of thin air, might have been for this exact reason.

Whether Prince was making a declaration of his true identity, or just singing lyrics he wrote for the song, Mike would have to investigate. One thing was for sure and that was that Brian would not be getting the album and poster. Instead, Mike gifted Brian with something the young boy appreciated much more. A prized shooting game Michael had bought at an estate sale where an electronic rifle would shoot down targets projected on a wall, or screen.

Photo of Mike prior to his attending a Prince concert upon being converted, like many others, as a dedicated Prince fan.

The next day Mike, began his research on Prince and was pleasantly surprised to learn that not only was Prince born on the seventh day (a prerequisite for anyone claiming to be the Messiah of which there had already been a few applicants who were of dubious character), but that he was born in the sixth month that actually aligned with the numerical alignment of 1989; at 6:17 pm in the sixth month on the seventh day in the 89th year of the 20th century. In fact, the number 7 followed Prince in ways that only the Universe could have manufactured. Apparently, Prince was aware of this and even wrote songs about the Holy number 7, having inserted it in numerous instances in his second movie, *Graffiti Bridge*. Despite his obvious overt sexuality, Prince was very much a spiritual fellow, which showed up in his lyrics and his charity work.

To aid his research, Michael began collecting magazines that featured Prince on the cover and eventually the evidence gathered was so overwhelming, he became convinced that Prince was indeed the reincarnation of the Apostle Paul. It mattered not to Michael that he might be the only person on planet Earth to suspect this since his own metaphysical background had already placed his philosophy so far outside the norm of what everyone else believed. This was also due to Michael's upbringing that he was able to consider a black man as the fulfiller of this important role.

Mike's Dad Jim had stressed by example that you judge someone by the content of their character and not the color of their skin, so when Prince popped up as a possible candidate, all that mattered was the evidence.

The problem now for Michael was how to express to Prince that there was indeed a witness to his true identity. It seemed imperative to Mike that Prince should receive some sort of confirmation and yet Michael had no credentials or noted accomplishments that might make his proclamation appear more credible.

While Mike pondered his lack of notoriety, a strange thing happened. The very same year he made the connection between Prince and prophecy in 1984, his hometown baseball team, the *San Diego Padres*, made it to the big show; the World Series in 1984. As every street corner celebrated the accomplishment, Michael suspected deep-down that this was a reward from the Universe for his courage in making his unorthodox assumption concerning the diminutive rock star. Neither was it beyond his comprehension that such a feat was orchestrated because he was unknown. His mantra was 'in the vast infinite scope of consciousness, all is possible', and ten years later, the same thing happened with the *San Diego Chargers* after a notable one-hour radio interview that Mike undertook with the host, Shelly on radio station KGBs *Public Market* radio show.

By this time, Mike had already manifested the means that qualified him to approach Prince and while he mostly kept his controversial Prince theory to himself, Mike had attracted the attention of *Public Market* as a result of his already printed front page UFO and rock face discoveries. At one point in the interview Shelly, cornered Mike after a comment he made that "he would gladly trade ten Super Bowl victories for the saving of mankind" when referring to the overdue shift of the poles. He then inferred that the Messiah was alive, so she asked him who it was and he refused to name Prince, but told the audience that he would leave some clues with Shelly for anyone who called in.

Mike told Shelly it was a famous celebrity, who had fulfilled prophecy according to metaphysics, and his birthday matched the numerical alignment of 1989. After the hour long interview the station was flooded with phone calls.

The very next football season, the *San Diego Chargers* made it to the Super Bowl in 1994 for the very first time in their playing history, which is a major accomplishment as this is the annual championship game of America's National Football League. However, it took several miracles for the *Chargers* to make it to the 'big game' and that clearly proved to Michael that some surreptitious manipulation was once again under way.

With three must win games to go in the season, Michael made an effort to help them achieve their destiny by traveling to a meet and greet with star Running Back, Natrone Means, at a Target store in Chula Vista, California. Mike gave a brief presentation to Natrone and the *Chargers* Defensive Lineman, Chris Mims who was also present and then handed him a copy of the *Public Market* radio interview. In return, Natrone gave Mike a hand-signed photograph.

The next week, the *Chargers* were about to lose to the *Kansas City Chiefs*, who drove the ball down the field behind, their talented Quarterback, Joe Montana. However, time ran out as Joe desperately tried to spike the ball down, to stop the clock and kick a winning 'chip-shot' field goal, resulting in a 20 to 6 win for the *Chargers*. In the very next game played by the *Chargers*, the *Miami Dolphins* were about to kick the winning field goal, which would have crushed any *Chargers* chance to play in the Super Bowl. With seconds left on the clock, and the *Chargers* barely in the lead, Michael physically had to remove a friend from his apartment who was bad mouthing the *Chargers*. Mike knew the Universe was working hard to help him and his team win, so there was no room for negativity.

Having cleared the room, Mike sat in front of the TV and yelled at the Miami Center to screw up the snap to the field goal holder, which is exactly what happened. It threw the kicker off and he sliced the ball to the right, missing the field goal. The next game against the powerful *Pittsburg Steelers* came down to the last play and as the *Steeler* quarterback threw what would have been the winning touchdown, fate tapped a *Charger* linebacker on the shoulder, who then turned to deflect the ball, with the final score in favor of the *Chargers*, 37 to 34.

Following the *San Francisco 49ers* Super Bowl win against the *Chargers*, Mike was about to throw away all the newspaper clippings he had saved, but then remembered what he had said on Shelly's radio show as having been willing to sacrifice ten Super Bowl rings for the actual saving of mankind. As a consequence, he retained these clippings, which decades later he still retains, to remind him of his need to persevere and advance in his own 'game', being the continued advancement of mankind.

Not long after Prince would have received Mike's package of pictures and news clippings at Paisley Park Studios in Minnesota, a powerful confirmation of Mike's Prince/ Messiah theory was aired on national television. On the popular TV show, *A Current Affair*, the host, Maureen, interviewed one of Prince's ex-girlfriends. After Prince had been described in glowing terms, his girlfriend remarked that Prince had themed bedrooms at Paisley Park, for his intimate rendezvousing pleasure and that "Prince thinks he's the Messiah". Michael was floored when she said this and felt relieved that he had indeed been onto something important, with those words having validated all his hard work and research.

It was with enormous dismay that legions of Prince fans awoke on April 22, 2016, to discover that Prince no longer occupied his physical body. He had died, apparently of an overdose of a potent pain medication known as Fentanyl being an opioid medication that Prince had been taking for years to relieve intense pain from troubled hip joints, following years of jumping off stages in high heels. Michael realized, as a chosen witness to the reincarnation of the Apostle Paul in the form of this musical genius, that he would have to post a tribute to him communicated on a popular social media networking site called *Facebook*. Mike had already completed dozens of radio interviews, decades earlier concerning his Prince theory and while *Facebook* would be full of people who knew him, he was already way beyond worrying about what family or friends thought of his oftentimes controversial and theoretical views. After all, he had been following a different drummer for years.

Those who knew Mike would also know that he speaks from the heart and truly believes in what he preaches, by backing it up with provenance in the form of published articles and evidence described as "Unsettling"[37]. He was also not fond of alcohol and had not been intoxicated since his early football days at Grossmont College.

In Mike's *Facebook* post concerning Prince, he included the same photo he sent Prince in the 1990s, which was a simple photo of dozens of magazines with Prince on the cover. It conveyed the sense that Prince was an accomplished artist who had won the highest accolades of the general public.

A selection of Mike's magazines on Prince, having become a 'true' fan!

37 Tony Perry, above n 32.

Mike was not surprised after making this *Facebook* post, to have only received a few positive comments. He had previously sampled his theory among friends and the idea was clearly outside their limited belief system. Besides, they already knew from Michael's previous postings that nothing was off-limits with this guy when it came to spiritual revelations. Mike had for quite some time been posting a variety of what may very well be seen as odd announcements concerning the paranormal and his discovering of the world's largest natural rock face on El Cajon Mountain, to that of his important front page UFO discoveries.

Basically, Michael had no fear about posting controversial theories and worried not that his reputation might be compromised, by proclaiming Prince as the Messiah. Mike retains the suspicion that Prince knew he was leaving this world and may purposefully have hastened his transformation, not only because of the unbearable pain he was in, but because he knew he was only two months away from turning 58 and perhaps preferred his last earthly age to be that of 57. As Prince was born on the 7th day, which is a holy number that permeates every area of our existence, so it was that we had 57 wonderful years with this dynamic individual, who not only moved millions of humans with his catchy tunes and spiritual message, but also represented for Michael at least, the closest thing to a modern day Jesus that we could have hoped for.

No doubt Prince is at this moment helping, in his own way, to find a better reality for the people he loved; being all of us. After all, Prince had already summed his life up in a message through these words, '*Love is the Masterplan*'.

In the mid-eighties, at the height of his interest in this Prince revelation, Michael was moved to make a very bold request to San Quentin State Prison in San Francisco. A local San Diego man, Robert Alton Harris, was about to be executed for killing two boys he chanced upon in Tierra Santa. The personal drama Harris suffered as a kid, being abandoned by the roadside by his mother at 13 years of age, led Michael to believe that Harris would not have committed the crime, had he not learned hate as the result of his broken childhood; a similar issue that has probably filled up our prison system across the country.

Mike's Dad was a great humanist, having turned their garage into living quarters for a homeless man who eventually landed a job and got his own place. Michael identified that the same man who killed those two boys decades earlier had morphed into someone completely different and here was good Americans willfully killing this man when God wasn't done with him.

Isn't the official motto of the United States of America, 'In God We Trust'? So, leave any revenge to a higher authority! Wasn't the Apostle Simon guilty of persecuting countless Christians before being converted? So, Michael boldly contacted Warden Daniel Vasquez and talked his way into sending them a fax that basically laid out his theory about Prince and how if he really is the Messiah then let him decide the fate of Harris. Mike knew it was really 'off the wall' to contact them and that it wouldn't make a difference, but the Universe was watching and he felt the spirit move him to make such an attempt. Ultimately, Robert Alton Harris merely represents many thousands of men and women who have lost their way, but are still redeemable with their destiny best left in the hands of God, and not with those making representation on behalf of the human race.

Mike's fax included numerous references to the number 7 and how it pointed to Prince as the elected one. Mike wondered if anyone who read his fax made the connection after Harris was killed, and how long it took for him to die in the gas chamber. According to the *Los Angeles Times* it took, 7 minutes for Harris to die, with his final stay of execution overturned by Federal Court Justices, delivering their verdict, 7 to 2.[38]

All this Prince stuff was summarized in a few sentences when Mike faxed the publisher of *SLAMM*, who had just written a glowing piece on Prince. Donavan seemed interested in both of Mike's theories on Prince and UFOs and sure enough, the following issue of *SLAMM* featured Mike's UFO story with a wonderful illustration of UFO's and a headline that read 'It's Like Martians Man'.[39]

Mike's article received top billing over stories on Burt Reynolds and G. Gordon Liddy. It also mentioned Mike's presentation to the Russian delegation in Rancho Santa Fe, California, that he had attended the week prior to this commentary being published.

---

38 Dan Morain and Tom Gorman, 'Harris Dies After Judicial Duel: 4 Stays Quashed', *Los Angeles Times*, April 22, 1992, <http://articles.latimes.com/1992-04-22/news/mn-507_1_death-penalty>.

39 Siggy Whitley, 'It's Like Martians Man', *SLAMM: San Diego's Lifestyle and Music Magazine*, (September 1994).

THEPRESSROOM

IT'S LIKE, MARTIANS, MAN

'It's Like Martians Man' as it appeared in the November, 1994, edition of *SLAMM: San Diego's Lifestyle and Music Magazine*.

Three days after thousands of copies of *SLAMM* hit the newsstands with Mike's UFO story featured inside, a wondrous thing occurred. Michael was watching *Channel 10 News* when anchors Kimberly Hunt and Carol LeBeau announced that three ground witnesses had called *Channel 10 News* to report how a UFO was trailing a *Channel 10 News* helicopter. Since this sighting happened so soon after Mike's featured article was published, he speculated that the sighting was undoubtedly the result of thousands of San Diegan's and their positive thoughts about UFOs, following their reading of his narrative in that issue of *SLAMM*.

One of the top teachings of metaphysics is that thoughts actually do have substance and strong thoughts result in physical manifestations, which is why Michael had been practicing the mental skill of mindfulness and 'right thinking' for decades, as this helps straighten the path ahead for those who follow this art form. One day the hidden power of thoughts will be widely known and practiced, and mankind will change his reality, merely by wishing it so. This theory showed up again in Mike's next major published article appearing in May of 1995.

Michael was given front page coverage in *GOOD TIMES Magazine* (unconnected to *High Times Magazine*) for this particular piece of writing.[40]

It was a huge piece with amazing photographs of Mike's *Famous Inaja UFO Photo*, along with two of the rock faces he had discovered, being the rock face in La Jolla Cove that had been featured two years earlier in the *Beach & Bay Press* and the largest natural rock face in the world, which Mike had called 'Orre' and had captivated thousands of El Cajon residents when this gigantic, yet stern looking face clearly depicted on the front of El Cajon Mountain, graced the front page of *The Daily California*, two years earlier.

The two journalists from *GOOD TIMES Magazine* wanted to include the rock faces in their article and Michael was only too happy to drive them to the site himself. Over 1,500 words discussed most of Mike's theories, but the most important story in that issue was concerned with the UFO encounter the Magazines' photographer (Johnny), actually experienced 2½ months prior to Mike's featured story. Johnny and his girlfriend came upon two large, dark orbs hovering directly in front of them, while traveling through Death Valley on Highway 127 at 11:00 at night. The objects left quickly and while Johnny admitted how frightened he and his girlfriend were, he affirmed how he looked forward to it happening again.

40 C. R. Davalos, 'UFORIA: Mike Orrell Says Aliens are Coming to Earth and They're Here to Help', *GOOD TIMES Magazine*, (May 31, 1995).

GOOD TIMES magazine May 31, 1995, front page of Mike introducing his critique on aliens coming to Earth and their mission to help.

Once again, Mike was secretly thinking that the emissaries 'saw', via the Akashic Records, that *GOOD TIMES Magazine* was about to run his chronicle of discovery in their magazine and they, the aliens, purposefully sought out the photographer who would be assigned to Mike's story, so that he would have some positive input in helping to encourage and complete the feature story in a big way.

Michael was delighted to hear from his friends how his face was all over the campus of numerous colleges when that issue of *GOOD*

TIMES Magazine hit the newsstands. For Mike, it was another confirmation that he was on a path with his heart in the right place.

Even though it would be years before his next feature article appeared in print, Mike would never give up on his quest to create a new perception of aliens. He firmly believed that aliens should be observed as benefactors, with an updated and realistic view of them thus driving the collective consciousness of mankind into the psychic existence required for humanity to qualify toward interplanetary contact and trade.

22

Research Persists

Not long after the *Science Fiction Channel* was created in the early 1990s, Mike decided to contact Dr. Franklin Ruehl, who was the host of one of Sci-Fi's new shows, *Mysteries From Beyond the Other Dominion*. Michael had created two amateur VHS videos, which contained much of the evidence he had gathered linking his *Famous Inaja UFO Photo* to other UFO photos taken around the world, as well as countless ancient artifacts. Mike convinced Dr. Ruehl to allow him to send his two videos entitled 'UFOs in San Diego', in an effort to have Dr. Ruehl include some of Mike's discoveries in his entertaining show.

In time to come, Mike was informed that three episodes of *Mysteries From Beyond the Other Dominion* would include Mike's discoveries, including the giant mile-wide natural rock face he had discovered on El Cajon Mountain. Unfortunately, just prior to these episodes being aired, Dr. Ruehl's show was canceled.

Two months later, Dr. Ruehl contacted Mike and asked him to send a copy of his *Famous Inaja UFO Photo* to the Senior Editor of the *National Examiner*, John Garton, who would be the one to decide if this magazine would run Mike's story. Mike sent through his photos and chronicle concerning his findings, with Mr. Garton being impressed enough to feature Mike's *Famous Inaja UFO Photo* and story in the magazine, placed directly above the credit box of the Magazines' writers. Mike was thrilled that his breakthrough UFO discovery had finally made it in a nationally distributed magazine, with the article including a direct quote from Michael that summed up the significant discovery he had made, being "one even had a distinctive, spike-like projection that was captured on three other UFO photos I've examined."

MYSTERY TRIANGLES ARE ALIEN CRAFT, SAY EXPERTS

Incredible photos reveal UFOs in flight

NINE UFOs fly in formation near San Diego

A FLEET of nine triangular-shaped UFOs has been captured in an amazing series of color photos shot by an amateur photographer.

The astounding pictures, taken near San Diego, California, by Michael Orrell, have UFO experts convinced they represent the overflight of extraterrestrial spacecraft.

By DR. FRANKLIN RUEHL

"These extraordinary photos are most persuasive evidence of the existence of UFOs," declares Delmer Nagle, a field researcher with the Intercontinental UFO Alliance, in Greenwich, England.

"What's so exciting is the distinctive triangular shape of these UFOs, nearly identical to UFOs I personally witnessed in June 1974 as they overflew the Ethiopian city of Asmara."

Orrell, of Pacific Beach, California, says he shot his pictures in July of [illegible].

He recalls: "Neither I nor my friends noticed anything peculiar in the sky. But when I developed the photos, I got the surprise of my life."

The prints revealed nine UFOs flying in formation. The 36-year-old account executive quickly had them enlarged.

"That's when I discovered the objects were triangular in shape. I'm convinced these UFOs are definitely not birds, planes or optical illusions.

"One even had a distinctive, spike-like projection that was captured on three other UFO photos I've examined."

These included one taken by astronaut James McDevitt during a Project Gemini flight over Hawaii on June 5, 1965, one taken by Vancouver, Canada, UFO researcher Dorothy Izatt and a third released by Russian cosmonaut Marina Popovich.

Article by Dr. Franklin Ruehl, as it appeared in the *National Examiner*.

With Michael continuing his valiant struggle to gain worldwide recognition for what he felt was the key to change mankind's future, he couldn't help but notice the planetary changes that were harbingers of what was appearing to be the fulfillment of Biblical prophecy concerning the end of days.

These warning signs included melting ice caps that not could contribute toward the raising of sea levels, with these immense icebergs possibly caving-in and changing the weight of the planet as they drifted into sea lanes, thereby causing an overdue shift of our already unstable planet. Just as these giant icebergs themselves often flip over when they become top heavy, was it also possible that they could cause our planet to do the same?

Mike never forgot the frightening vision of the Indian shaman who was interviewed on TV decades earlier and claimed to have had a vision of all of Earth's trees flying through the air. That nightmarish vision meant nothing to the interviewer, but to Mike, it made perfect sense. After all, should the Earth physically flip over, with geophysics referring to such an event as a crustal displacement, 500 mph winds would be produced that would not only send trees flying, but every structure on the surface of the planet.

Revelation 6:12-14 accurately depicts the devastation that the shifting poles will unleash:

"And I behold when he had opened the sixth seal, and, lo, there was a great earthquake; and the sun became black as sackcloth of hair, and the moon became as blood;

"And the stars of heaven fell unto the earth, even as a fig tree casteth her untimely figs, when she is shaken of a mighty wind.

"And the heaven departed as a scroll when it is rolled together; and every mountain and island were moved out of their places." [41]

---

41 Revelation 6:12-14, King James Version.

Edgar Cayce had practically the same thing to say when describing the long and overdue shift of earth's magnetic poles, as "What is the coastline now of many a land will be the bed of the ocean. The greater part of Japan must go into the sea. The upper portion of Europe will be changed as in the twinkling of an eye. There will be a shifting of the poles so that where there have been those of a frigid or semi-tropical will become the more tropical".[42]

Michael had suspected for decades that this is exactly what happened to dozens of flash frozen mammoths who were found in the frozen wastelands of the arctic with bellies full of summer food. The revelations account in the *Bible* specifically mentioned 'a strong wind', which is what is predicted to happen as the earth's crust rapidly slides around the core like the outer peel of an orange. Mike imagined if he were standing on the earth's surface and looking at the stars in heaven as such a shift happened, the stars would indeed appear "as a scroll when it is rolled together".[43]

In the back of his mind, Mike was certain that everything happening now was constructed without openly being observed, so that mankind could somehow escape this most devastating of catastrophes, unlike anything ever experienced before. He was convinced that courageous alien lives had even been sacrificed in numerous UFO crashes around the globe so that mankind could speed-up its technological advancement by back-engineering alien technology.

This is also the theory of ancient alien theorists and one that aliens have planned all along. Mankind would be a space traveling race and only recently have we been capable of utilizing and harnessing the technology that these numerous crashed UFOs and dead aliens have been providing.

---

42 William Hutton and Jonathan Eagle, *Earth's Catastrophic Past and Future: A Scientific Analysis of Information Channeled by Edgar Cayce* (Universal Publishers, 2004) 175.

43 Above n 41.

For the sake of these hero pilots from planets unknown, Michael vowed to never give up in his pursuit to spread the word of the benevolent nature of their existence and how mankind needs to become humble and fall in line with the alien plan for our future redemption.

In May of 2006, Michael got the call he had been waiting for from *KFMB-TV Channel 8*'s Senior Producer, who asked Mike if he was interested in being interviewed for a piece concerning his UFO theories. Of course, he accepted and asked when they would like to see him at their studio? They replied that they would prefer to film at his apartment, which Mike agreed too.

Following this conversation, Michael realized that *KFMB-TV Channel 8* was also interested to see if he was a UFO fanatic, with model UFOs hanging from the ceiling and *X-File* posters on the wall. He had already encountered such a hidden agenda years earlier when *GOOD TIMES Magazine* sent their investigative team to Mike and Ronnie's apartment in Pacific Beach to carry out their story, before they headed to two different rock face locations, being El Cajon Mountain and La Jolla Caves. Mike is, unfortunately, a bit of a hoarder, especially when it comes to holding onto anything having to do with his research on UFOs as he has hundreds of books, boxes of newspaper articles, papers, faxes, and artwork. None of this clutter seemed to bother him though and his philosophy, for anybody who inquired, was that history only cares about his discoveries and not his cleanliness.

Prior to Mike's upcoming TV interview with KFMB-TV Channel 8, he decided to try and find a local photo developing business to further enlarge the negative of his *Famous Inaja UFO Photo*. The first photographic store he tried on University Avenue in San Diego failed miserably to enlarge the objects to Mike's specifications. He took the negative to another place up the street, but as he approached the female attendant at the counter of the second store, Mike found that his negative was missing from its folder.

Mike was in awe of this negative that was destined to change the world's reality, yet he had just lost it. He excused himself and followed his footsteps back to his then Saturn motor vehicle and after an exhaustive yet unsuccessful search in his car for this 'fucking' negative, he decided to race back to the first miserable store he had attended, to see if he could pull out a miracle.

Upon checking the parking lot, Mike saw no sign of his negative and offered a reward to the owner of the store, should anyone find the negative lost by Mike, in his deluded and quite a hysterical state, having begun to actually suspect the store owner of stealing. Mike then raced back to the second parking lot, with his chin dragging on the pavement, whilst beseeching God to produce the negative should he want for Mike's story to be told.

Mike again walked back into the second photographic business to see if someone had miraculously turned it in, but no-one had. As he stumbled out the door, grief-stricken, he noticed his negative still in its paper holder, sitting on the ground in the middle of the parking lot. He knew that he had just walked by that way and it had not been there. After closely examining the negative, he breathed a sigh of relief as its condition remained pristine. Mike was thankful for the second-chance mulligan he had been given and he promised the Universe that he would be extra careful in the future.

When the *KFMB-TV Channel 8* Photo Journalist and Senior Producer showed up at Mike's apartment complex, he was ready with a semi-clean, but still clustered and near borderline hoarding apartment located in Redwood Village near San Diego State University. What these news reporters were looking for, was any sign that Mike was more on the lunatic fringe of this popular world topic of UFOs. What they saw; however, was Mike's fine art and scenic photography hanging on the walls and not a single shred of evidence that this guy was anything but a normal guy (although with hoarding tendencies), who one fortuitous day took an unlikely photo of UFOs.

With his girlfriend, Ronnie, waiting in the bedroom, the *KFMB-TV Channel 8* camera rolled in their living room, as Mike went into presentation mode and repeated what he had said in dozens of radio interviews across the State, which was that these patterns are significant and cannot be ignored.

Mike spoke from his heart, which is the 'warrior for knowledge' way and he dared not come across as 'preaching' since he was most comfortable playing the role of a student and amateur UFOlogist. After all, a wise man knows just how much he actually does not know!

The interview went well and Mike was surprised and pleased when afterward they asked him to lead them up to Santa Ysabel and continue the interview at the actual spot he had filmed the UFOs, way back in July of 1990. They also wanted to see and film the alien boulder that was featured in *The Daily California* in 1992 which is just feet away from the platform where Mike photographed the ten hovering objects from another world. Mike agreed to lead them all the way to Inaja Memorial Campground and soon the *KFMB-TV Channel 8* News van was following Mike in his Saturn, as they made the scenic drive through Lakeside, Ramona and eventually Santa Ysabel.

Mike made sure to pull into Dudley's Bakery, which was located a few miles west of the entrance to Inaja Memorial campground in Santa Ysabel. He explained to the two journalists that he always stopped there for fresh fruit bars, such as he had that fateful day back in 1990 on the taking of his *Famous Inaja UFO Photo*. Once the two journalists had refreshed themselves, they continued driving up the hill, with all being quick to arrive at the entrance to Inaja Memorial Park.

The news crew grabbed their gear and started on the path, leading firstly to the memorial marker erected at the Park's entry. Mike took a moment to explain the coincidence of how he was born one day before the fatal fire that killed 11 men, which why the Park was constructed. The television crew took some footage of the entrance and moved on.

After traversing what is clearly a well-designed scenic trail, the trio at last arrived at the real treasure; a majestic view overlooking 18 miles of the San Diego river valley at Scenic Spot 7. Even before they reached the platform, the two journalists discovered the alien shaped boulder for themselves and were visibly pleased. They had Mike again traverse the trail and walk up to this mysterious alien head boulder while filming him. When Mike reached the boulder he did what he always does when he makes such a pilgrimage to this sacred spot; he wrapped his arm around the magical rock and thanked the Universe for his many blessings and for orchestrating a cosmic event at this very spot decades ago. They also filmed him on the same platform where he had taken his *Famous Inaja UFO Photo*, whilst discussing the events that led to him taking this photo. A few more shots and it was time for them to leave.

To their horror and amazement, as the three hiked to a ridge overlooking the parking lot, they discovered that the KFMB van's side door had been flung wide open. Mike was worried that this would be a bad omen, should thousands of dollars in camera equipment be stolen while they merrily filmed away elsewhere. While the two journalists scrambled down the rest of the trail, Mike kept an eye on the van and was pleased to see no-one coming out of it. At last, the crew arrived at the van and breathed a sigh of relief, noting that nothing had been disturbed.

As Michael finished the trail and joined the journalists in the parking lot, he mentally thanked the Universe for protecting their van. They said their goodbyes and made their individual ways back home. Michael was confident that the news team would edit a strong piece on his behalf and that he had also given this interview his best shot, come what may. It wasn't until he arrived home and looked in the mirror that he realized he had buttoned his shirt incorrectly, skipping one of the loops when he hastened to change it from what he had been wearing when interviewed at his apartment. This blunder left one-half of the front tail of his shirt longer than the other, with his collar also pulled down on one side. It was an inexcusable mistake that bothered him for days afterward.

It was when *KFMB-TV Channel 8* started airing promotions of the upcoming special on Mike's UFO story that he forgot about his wardrobe malfunction and concentrated on spreading the news of his upcoming paranormal special. His technique of harvesting potential contacts from any related articles in the newspaper or science magazines had resulted in numerous front page articles and even this TV interview. He would literally spend hours surfing the Internet to potentially find important email addresses to send his press release too, which has since become an endless task.

The creation of the World Wide Web allowed for Michael to easily share the sacred patterns he believes will change the course of humanity. He had to spend a lot of money, in the beginning, faxing information to his chosen targets and watched with satisfaction as the personal computer changed our world and his.

The night had finally arrived and as the *KFMB-TV Channel 8* piece aired, Mike and his girlfriend roared with laughter at the mere thought that he was on TV. Mike's favorite spot in the interview happened when they took his 35 mm negative and had photo experts authenticate that the objects were indeed ingrained within the negative and that it was untouched and not damaged in any way. In the background, they played space music, such as that from the *Twilight Zone* after having concluded that the objects were unidentifiable. Which is a miracle considering what had happened to the UFO negative as Mike was preparing for the interview.

Mike received a few congratulatory phone calls right after the interview and several back slaps at the gym. One of his friends, Zippy, commented that the part of the interview where Mike hugged the alien head boulder and listened for wisdom, as he cupped his ear to the alien boulders mouth, was hysterical. Apparently not everybody thought so.

Mike's Mom was upset that he did not appear professional, as after all, he was listening to a rock! She also said that the two female anchors laughed at the final scene and in his Mom's eyes this was insulting. Mike explained that it shows he has a sense of humor and that he does not take himself as viewed on television so seriously. Besides, he's not a professional ... just look how sloppily he's dressed! Mike then shamefully pointed out to his Mom, "My shirt is clearly mis-buttoned", he moaned. She laughed and said that she had not even noticed.

Image of Mike as shown during his *KFMB-TV Channel 8* interview aired in May of 2006.

Years later those two female anchors paid the price for their innocent but sarcastic snickering when they were edited out of Mikes UFO interview, which would eventually win first place at the '4th Annual McMinnville UFO Film Festival' held in Oregon in 2014. This *KFMB-TV Channel 8* interview was broadcast in 2006, which means Mike's eight-year-old interview was still relevant and to a large group of like-minded people, it most likely represented the closest thing to disclosure that civilians are going to get. After all, the *KFMB-TV Channel 8* segment actually included a montage of some of the sacred patterns that Mike discovered, thanks to his *Famous Inaja UFO Photo*, putting them into the history books and earning Mike's undying appreciation.

Mike's historic interview was the catalyst to finally win over a 21 year campaign to be featured in this county's largest regional newspaper, being *The San Diego Union-Tribune.*[44] Mike received a call from their senior writer, who asked if he would like to be interviewed for a piece in *The San Diego Union-Tribune*, which of course, he was.

Mike Orrell, with an enlargement of his UFO picture taken in 1990 that appeared in the *San Diego Union-Tribune* on July 5, 2011 (photo credit Nelvin C. Cepeda).

Mike offered to show up at their facility in Mission Valley, but the writer insisted he come to Mike's own home, accompanied by a photojournalist. Mike agreed, knowing that just like the *KFMB-TV Channel 8* team, these guys needed to get a feel for their subject matter. No doubt they wanted to know whether this guy was a 'kook', who was just scamming people with a bogus UFO story for 15 minutes of fame, or had he really made an important discovery worthy of front page coverage? As it turned out Mike and his evidence for extraterrestrials was worthy of front page coverage and the positive and lengthy article, including photos, launched another wave of congratulatory phone calls and catcalls.

44 John Wilkens, 'UFO Believer Tries to Spread the Word', *The San Diego Union-Tribune* (online), July, 5, 2011, <http://www.sandiegouniontribune.com/sdut-ufo-believer-tries-to-spread-the-word-2011jul04-story.html>.

At the conclusion of Mike's *San Diego Union-Tribune* interview, he gave John Wilkens, the senior writer, copies of the enlargements of the Inaja objects and challenged him to have the *San Diego Union-Tribune's* photographers examine the objects to decide for themselves and come to their own conclusions, which they apparently did.

Mike's photo had previously been examined by *CBS News 8*'s partner, *KFMB-TV Channel 8*, who had taken his original negative to Chrome Photo, based in Sorrento Messa, California. Upon scrutinizing the negative and having been asked for their opinion, they reported that the objects within the printed photograph were certainly irregularly shaped and appeared to be triangular in form. On even closer inspection, staff at Chrome Photo completely ruled out any dust as being on the lens, or damage incurred on the negative itself. No mechanical defects were detected with Mike's camera that took this photograph.

"The negative itself was intact," staff member Dennis Reiter said. "There was no damage, no machine type damage in the process."[45] Chrome Photo continued to scrutinize Mike's photo by placing his negative through a high-resolution scanner for it to be better viewed on a computer screen. When asked whether, in their opinion, what appeared in the photo may, in fact, be UFO's, Reiter added, "I would agree with him - they are flying objects and they are unidentified", thus confirming the authenticity of Mike's accidentally taken photo depicting ten daylight objects clearly labeled as 'Unidentified Flying Objects'.[46]

Mike's *San Diego Union-Tribune* article was delayed from being published a couple times because they wanted to fit it on their front page, which they eventually did. It was a great coup for Mike

---

45 UFO Casebook: UFO Sightings, Case Files, Photos, Videos, Alien Abduction, *Man Says Photo Shows UFOs Over Santa Isabel*, (May, 11, 2006), <http://www.ufocasebook.com/inaja.html>.

46 Ibid.

to have the largest newspaper in regional San Diego, write a positive piece about such a controversial subject. He then worked overtime on his computer to spread the news about what was this latest feather in his UFO cap.

Not long after Mike's article was published that July 4th of 2011, the senior writer contacted him as a direct result of having been contacted by so many people calling the *San Diego Union-Tribune* with their own UFO stories; similar to what always seemed to occur after the publishing of any of Mike's other articles. He was given the contact numbers of those who had called, with one caller having a story that was truly spell-binding.

This individual worked in the scientific field as his conversation reflected. He lived in the East County, near Cuyamaca Lake, and he detailed an event that happened within 2010 when he and a male friend were climbing the steps to a restaurant located across from the lake. One of them noticed three lights hovering over the south end of Cuyamaca Lake and as the both watched with interest, they suddenly realized that there was a huge, hovering, triangular craft attached to these lights. Almost instantly, they found themselves aboard the craft and were subjected to the same physical examination that most abductees maintain as happening to them.

As Mike questioned this man, he claimed that he and his friend were returned to the same spot on the stairway where they were initially taken, and that they had noticeably lost almost an hour. This witness alleges that his companion still refuses to discuss the event years later after it occurred, as he was and remains terrified. Michael inquired as to whether or not they had been hypnotized, such as claimed by Betty and Barney Hill during their own alleged abduction, way back in 1964.[47]

Mike's witness stated that he had been in contact with a qualified female Regression Therapist, to assist with the performing of any reversion back to when he experienced this incident, but his fellow abductee absolutely refused to participate. Mike reassured the witness that, as terrifying as his experience was, it remained Mike's belief that aliens have our best interest in mind and that we are a direct product of their own intervention, and have been since the beginning of time.

The one thing that Michael came away with after this interesting conversation is that San Diego seems to be some kind of gathering place for aliens and their craft. Besides this giant, triangular UFO encounter experienced by these two serious-minded men in the region of Cuyamaca, San Diego, back in 2010, there were a further three major sightings of UFO fleets, which were captured on video and 35mm film, including that of Mike's own *Famous Inaja UFO Photo.*

Fox 6 News aired a remarkable video of nine UFOs in, complete formation filmed by a person attending a 2008 New Year's Eve party in Claremont, San Diego. Mike was excited that these objects appeared to resemble his own fleet of UFOs and were not Chinese flaming lanterns, as some of his skeptics had already suggested.

There was also the incident recorded by a cameraman from *KNSD-TV Channel 7*, who in 2014, captured a fleet of UFOs hovering over the ocean in Imperial Beach, California. In addition, a former News Anchor, Denise Yamada, undertook a special report on a particular type of UFO known as 'rods', when she accompanied

---

47 UFO Evidence, '*Betty & Barney Hill Abduction Case*' (2011) <http://www.ufoevidence.org/topics/bettyhillcase.htm>.

Jose Escamilla, an expert on capturing these huge lightening fast rods that zip through our atmosphere so fast that they are undetectable, unless videotaped on sport speed using a video camera. Sure enough, when they went to the top of Kate Sessions in Pacific Beach and filmed the sky, they later discovered upon review that they had indeed captured numerous rods cruising unseen in the skies above San Diego, and most likely around the world.

Mike wondered about the purpose of these rods. What is their mission? Are they gathering information from us by tapping into our transmissions or even our thoughts? Are they reporting back to a 'mothership', or some higher authority? In Mike's own *Famous Inaja UFO Photo* he had discovered two beams that enter the frame from the top and go all the way to the ground. The beam on the left actually intersects with the fourth UFO from the left, affecting its shape. The other beam is on the right side of the valley and is clearly leading the single craft, as it approaches the formation of the nine others evident on the left-hand side of Mike's *Famous Inaja UFO Photo.*

It seems clear to Michael that with this long history of UFO sightings in and around the great City of San Diego, which is also his fair city and destined to be ground zero for alien contact. With this gigantic human face sitting on the front of a local mountain, then perhaps 'they' plan a Steven Spielberg type entrance, such as that depicted within one of his scenes from the hit movie *E.T. the Extra-Terrestrial.* A huge mother-ship dropping down and hovering directly over the mysterious face on El Cajon Mountain, as scout ships buzz to and fro with their blue lights illuminating the entire mountain – now, this would be a sight to see!

23

Michael's Quest Continues

For years, Mike has symbolically held his discoveries over his head, while trudging through the muck that can be the daily grind of being human. He struggled to keep his research together, while changing girlfriends and addresses, knowing full well that past events would not care how he did it, but that he simply must continue to do it. One day, he is sure that history will catch up with him and he wants to be ready.

A large collection of rejection letters has not deterred Mike in his quest to bring his discoveries to the world's attention. He wants for everyone to closely scrutinize the evidence that he has presented within this book, so readers can make their own informed decisions and even contribute to his research, by way of contact and networking directly with him.

Mike's research continues, without any form of monetary support that is neither expected, nor requested. He actually considers himself fortunate, even lucky, to have found this lost alien code and the fact that he was the first human to stumble upon it means that no-one else may have been looking for what, to him, has become an obvious signature left behind proving that aliens not only exist, but are in fact responsible for our physical construction. All this, just to serve the desire of our 'Creator' or 'Godhead' and for everyone to have the opportunity to become a 'Godlet', so they may control their own gift of free will that holistically contributes to all mankind.

For Mike, it has been hard to ignore that he was chosen for the role of lead cheerleader, trying to paint a more realistic and positive portrayal of aliens to anyone who would listen, but especially those citizens in the San Diego area who were open-minded enough to weigh the evidence fairly and make the same conclusions that he had made, which is that alien contact is a good thing, and is in the best interest of the entire human population.

As unqualified as he is for this mission, Mike also knows that he has gone through doors now closed shut behind him. TV shows that provided critical evidence have been canceled and several publications where he was featured on their front cover lost their advertisement war and shut down. Mike is still determined not to let down those courageous journalists who boldly followed their instinct and published his story. He persists by using all resources available to him, along with his own determined courage to boldly go forth and cast his net of UFO knowledge across the Cyber Sea.

Mike continues to look for the fertile ground where his online evidence will convince an established producer or director to take an interest in this true paranormal tale and create a documentary that will shake the very foundations of our established sense of reality, which could literally change our world for the better. He is convinced that if his discovery was given to the general public at large in a professional film format that would circle the globe, his presented evidence would be so compelling and impressive that the entire civilization would re-align their perception of aliens from menacing man-eaters, to benevolent benefactors. As it stands, with mankind embroiled in wasteful pursuits not aligned with man's true purpose on earth, the aliens continue to wait patiently for us, their prodigy, to wake up to the real reality… a psychic reality.

A revolution of new thought would expand across the globe as the idea of actual interplanetary contact captures the imagination and awe of billions of humans, ready for a new way of life that more closely reflects the source from which we sprang, as the spirit created the flesh and not the other way around. It may happen in the wee hours of the morning, or they may arrive in the form of a majestic daylight arrival, but either way, it appears that any rapture-like event is on the horizon. With all the natural disasters and man's increasing violence toward other men, Mike feels that a day of reckoning is soon at hand when aliens from literally everywhere in the cosmos, will arrive to airlift mankind to safety, with a new paradise of non-violent humans and colony of off-worlders thus beginning.

Michael realized long ago that perfection is not expected, just a good attempt at searching for what is real, finding it, and following its course with only a passing interest in the temptations of the fleshy sack that one's divine consciousness inhabits. After all, everything we do, think, or say is being recorded in the Akashic Records, so we all are literally getting away with nothing.

One's karmic debt will be paid in either this lifetime or the next, but no one gets out of this schoolhouse without firstly qualifying. At least a worthy attempt at following *The Ten Commandments*, which we and every civilization has been charged with keeping, should be on the scorecard for everyone who cares enough to want to hop onto the alien bandwagon.

In the end, this is why the physical Universe was created in the first place; it really was just for us. A chance to use free will and discover that 'Love is the Masterplan' (Prince), and that it's okay to be part of a good thing. If the person in charge is doing a great job, then this should be good enough for any individual. Vanity and jealousy, along with selfishness, are negative emotions and detract from the beauty and love everywhere around us, which represents the source of our creation, and was the only inspiration for our creation by the Universe and/or our God.

Our existence would be free of disease, homelessness, starvation, and even death. The daily agenda would include a re-education for all survivors, beginning with a close look at what really happened in our past, so that we can dispense with the myths and legends and move forth with what really happened in our past, as after all, our past leads to our future and enables us to make informed decisions as human beings on what the best course for all of us to follow should be.

In place of the chaos and disharmony that currently persists all over our planet, and the greed and corruption over the struggle concerning the almighty dollar, there will instead be a renewed sense of purpose, as mankind adopts established trading practices of well-established galactic communities and allowing for a real golden age to commence here on Earth.

Right-minded humans who have proven their worthiness to be part of this next, evolutionary phase (interplanetary contact and trade), will rejoice as new generations are raised in comfort and security, knowing that at last humanity is on a path with a heart, and it is all because humanity willed it thus.

Michael wholeheartedly believes that the same sacred pattern he first discovered in his *Famous Inaja UFO Photo*, is also purposefully plastered on the face of the moon in the gigantic 'projecting acorn-shape' that is Mare Serenitas. Once these two matching patterns have been viewed and connected, it becomes impossible to not see this sacred shape on the surface of the moon. Just like in the movie *Dune*, when Paul Atreides took his new name of Paul Muad'Dib from the mouse shadow on a nearby moon, so too would mankind only need to look up at night to see the sacred symbol chosen to represent the emissaries from God, being extraterrestrials.

Mike holds no allusion concerning his place in all of this UFO discussion, or how his findings may be perceived, particularly by unbelievers. He is not an archaeologist, nor a professional researcher, or even a learned man. He is just an average person, who happened to be at the right place and at the right time to accidentally take his *Famous Inaja UFO Photo* that clearly shows objects visible within and to this day remain unidentified.

To have this regular man making these preposterous claims that he's actually solved a UFO mystery has always been 'a bit much' for people to digest. Mike's response remains that it is not about who found this lost and sacred pattern. It is about the pattern itself, being one that in time to come will enable extraterrestrial communication between Earth, and others, if only mankind has the intellect, vision, and desire to openly pursue, and research evidence as it comes to hand, such as presented within this book.

The substantiation of facts found by Mike and delineated within the pages of this book is certainly compelling, although with every rejection he has received, his resolve also strengthens. Mike regards it as his responsibility to persist with his UFO investigative studies and somehow get his story out. He is, after all, the first person to openly have had the courage of decoding his findings. This has become a daunting, but necessary task for this poor and private citizen, who for one reason or another has been charged with the duty of presenting his researched evidence to the public at large.

Mike persists in placing his trust in the Universe, so that he may carry on his efforts for the betterment of mankind as he remains resolute that one day, he will indeed succeed in his mission of helping to save his civilization. Undoubtedly, this still means that for Mike, he must juggle the requirements of daily life, while still trying to generate awareness of his findings.

Over the course of many years, Mike has completed 32 radio interviews, with several being over two hours long. He has been questioned, even interrogated, by most people he meets when conveying his theories and evidence gathered to support such thinking. However, the day may indeed come when aliens openly show themselves, which would support all free thinking people such as Mike, and many others who firmly accept as truth that aliens have been aiding mankind from Earth's very commencement. Mike is not the only person that thinks they deserve our love and gratitude, along with our obedience to whatever rules these ancient trading partners may choose to set up.

On balance, one day this will be mankind's reality and there will be no more sad endings, as this generation will be *The Last in Line* to have lived through such transformation.

Reference List

Bridges, Marilyn and Maria Reiche, *Markings: Aerial Views of Sacred Landscapes* (Arpeture, 1986) 9, 27

Caviezel, Jim, 'Passion' Star: 'You Have to Take Jesus', (August 10, 2014) World Net Daily (WND), <http://www.wnd.com/2014/08/passion-star-you-have-to-take-jesus/>

Congregation for the Doctrine of the Faith, *The Message of Fatima* <http://www.vatican.va/roman_curia/congregations/cfaith/documents/rc_con_cfaith_doc_20000626_message-fatima_en.html>

Davalos, C. R., 'UFORIA: Mike Orrell Says Aliens are Coming to Earth and They're Here to Help', *GOOD TIMES Magazine*, (May 31, 1995)

DeLeon, Clark, *Pennsylvania Curiosities* (Rowman & Littlefield, 4th ed, 2013) 131

Elling, Regina, 'INAJA Monument: A Tribute to the Fallen', *The Guide to Julian CALIFORNIA* (2010) <http://www.orreman7.com/JulianGuide.html>

Event Chronicle, '*6 Astonishing Irregularities That Are Evidence of a Hollow Moon Space Station*' (August 27, 2015), <http://www.theeventchronicle.com/metaphysics/galactic/6-astonishing-irregularities-that-are-evidence-of-a-hollow-moon-space-station/>

Fayette County Cultural Trust, *Keckburg UFO Incident*, <http://www.fayettetrust.org/Kecksburg-UFO-Incident.html>

Gordon, Stan, 'The Kecksburg, PA UFO Crash Incident', *UFO Evidence: Scientific Study of the UFO Phenomenon and the Search for Extraterrestrial Life* (2011) <http://www.ufoevidence.org/documents/doc1300.htm>

Harris, Paola, Exopolitics, *San Diego Sightings are equally important!* (2007), <http://www.paolaharris.com/sandiegostory.htm>

Hutton, William and Jonathan Eagle, *Earth's Catastrophic Past and Future: A Scientific Analysis of Information Channeled by Edgar Cayce* (Universal Publishers, 2004) 175

Kasser, Rodolphe, Marvin Meyer and Gregor Wurst (eds), *The Gospel of Judas* (The National Geographic Society, 2006)

Morain, Dan and Tom Gorman, 'Harris Dies After Judicial Duel: 4 Stays Quashed', *Los Angeles Times*, April 22, 1992, <http://articles.latimes.com/1992-04-22/news/mn-507_1_death-penalty>

NASA, '*Gemini IV*' (April 14, 2008) <https://www.nasa.gov/multimedia/imagegallery/image_feature_1061.html>.

New Awareness Network, *Seth: The Spiritual Teacher that Launched the New Age* (online) 2003 <http://www.sethlearningcenter.org/>

Orrell, Michael, *Best UFO Photo Ever 2: The Projection is the Rosetta Stone* <http://www.orreman7.com/BestUFOphotoever2.html>

Perry, Tony, 'Hey, You Think It's Easy Being Chosen to Spread the UFO Gospel?', *Los Angeles Times*, June 12, 1991

Ruhl, Don, *List of Sevens in the Bible* (online), October 12, 2016, <https://sevensinthebible.com/list-of-sevens-in-the-bible/>

Starship Earth: The Big Picture, *'NASA Forgot to Tell Us About the Transparent Dome and UFOs on the Moon in 1969'* (February 23, 2013), <http://www.starshipearththebigpicture.com/2013/02/23/nasa-forgot-to-tell-us-about-the-transparent-dome-and-ufos-on-the-moon-in-1969/>

Stewart, Gail Barbara, *The Bermuda Triangle* (Reference Point Press, 2009) 36

Sutherland, Billie, 'Making a Martian out of a Mountain', *The Daily Californian*, (May, 3-4, 1992)

Sutherland, Billie, 'Thar's Faces in Them There Hills, *The Daily Californian*, (May, 17, 1992)

UFO Casebook: UFO Sightings, Case Files, Photos, Videos, Alien Abduction, *Man Says Photo Shows UFOs Over Santa Isabel*, (May, 11, 2006), <http://www.ufocasebook.com/inaja.html>

UFO Evidence, *'Betty & Barney Hill Abduction Case'* (2011) <http://www.ufoevidence.org/topics/bettyhillcase.htm>

Whitley, Siggy, 'It's Like Martians Man', *SLAMM: San Diego's Lifestyle and Music Magazine*, (September 1994)

Wilkens, John, 'UFO Believer Tries to Spread the Word', *The San Diego Union-Tribune* (online), July, 5, 2011, <http://www.sandiegouniontribune.com/sdut-ufo-believer-tries-to-spread-the-word-2011jul04-story.html>

Wilmont, Mary 'Pacific Beach Resident Photographs UFOs in San Diego', *Beach & Bay Press*, April 23, 1992

CPSIA information can be obtained
at www.ICGtesting.com
Printed in the USA
FSOW02n0755231017
40225FS

9 781546 458395